Models of Biological Pattern Formation

Models of Biological Pattern Formation

Hans Meinhardt

Max-Planck Institut fur Virusforschung
Tübingen
West Germany

1982

ACADEMIC PRESS
A Subsidiary of Harcourt Brace Jovanovich, Publishers

London   New York
Paris   San Diego   San Francisco   São Paulo
Sydney   Tokyo   Toronto

ACADEMIC PRESS INC. (LONDON) LTD.
24/28 Oval Road
London NW1

United States Edition published by
ACADEMIC PRESS INC.
111 Fifth Avenue
New York, New York 10003

British Library Cataloguing in Publication Data

Meinhardt, Hans
    Models of biological pattern formation.
    1. Developmental biology—Mathematical models
    2. Biological patterns—Mathematical models
    I. Title
    574.3'0724   QH491

ISBN 0-12-488620-5

Filmset by Eta Services (Typesetters) Ltd., Beccles, Suffolk
Printed in Great Britain by
Thomson Litho Ltd, East Kilbride, Scotland

Preface

The question of how development is controlled is one of the challenging problems of biology today. Many fine experiments have been carried out on this topic. However, despite all this effort, the underlying mechanisms remain obscure. Many different interactions between molecules, cells and tissues must be involved in the process of development. Our intuition as to how such multicomponent systems behave can be very misleading. The understanding of other complex systems, ranging from problems in economics to engineering, has been greatly advanced by the use of precise mathematical models. The properties of these models are studied by analytical considerations as well as by computer simulations and are then refined by comparison with the properties of the real system. In this book, general classes of molecular interactions that lead to biological patterns are presented along with detailed modelling of particular developmental systems. It will be shown that a relatively simple set of interactions can explain seemingly complex experimental observations in a quantitative manner. I hope that these theories will provide a framework for further experimental investigations and create insights which will facilitate future biochemical studies. Discrepancies between these theories and these future experimental results will lead to modifications and refinements of the theories and, hence, will focus new experiments. At the present state of the art, theoretical explanations are necessarily hypothetical. They cannot deal with the enormous complexity of biological development, but rather with a set of general features. Even if we select a particular experimental system for detailed analysis and modelling, the selection is made by judging which features are typical for biological development as a whole and are thus a test for the scope of a general theory. It has to be stressed that the selection of the biological examples has a subjective component. The selection of examples chosen in this book has been made by considering whether they explore the potential of simple reaction schemes capable of pattern formation (not that the systems are necessarily always simple but simple schemes provide a good first approximation). This book deals with the problem of how cells in one part of an organism become different from those of other parts. The change of shape and form (the morphogenesis proper) which is thought to be a consequence of these primary differences, is not considered.

It is difficult to create a sequence in a book on developmental mechanisms. Ideas, much like developmental mechanisms, branch from one another like the limbs of a tree, but the sequence of a book is necessarily linear. The first section of the book describes mechanisms which have the ability to generate biological patterns. The following sections deal with how these mechanisms are used to generate positional information. A detailed comparison of the general theory with insect development is given. Later sections describe the generation of subpatterns, as well as how cells respond. Mechanisms for the activation of particular genes will be presented which are compatible with developmental alteration induced by a known class of mutants. The book will conclude with a collection of computer programs, so the reader can simulate the assumed interactions himself.

Many have contributed to the development of these theories. Most important is Dr Alfred Gierer, who provided the initial insight concerning the necessity of autocatalytic processes. Our fruitful collaboration over the years has been responsible for the interest and excitement my work has given me. The Max Planck Institute for Virus Research in Tübingen offered a fertile environment for these ideas to grow. My warmest thanks go to all my colleagues for their helpful discussions. I also thank Hans Bode, Richard Burt, Anthony Durston, Scott Frazer and Harry MacWilliams for their patient help in the preparation of the manuscript as well as Gunilla Weinraub and Aiko Tanaka-Ingold for the drawings. This book has grown out of a previous review article (*Rev. Physiol., Biochem. Pharmacol.* **80**, 48–104, 1978) and I gratefully acknowledge the permission of Springer Verlag to incorporate some of that material here.

May 1982   Hans Meinhardt

Contents

1

Theories as a necessary supplement to experimental observations

The control of development in a higher organism is one of the major unsolved problems in biology. How development has to proceed must be genetically determined. However, the genes themselves cannot directly explain the generation of differences in different parts of an organism since the genetic material in most cells of an organism can be assumed to be the same. Nor can biochemical investigations alone lead to a full understanding of development. Even an ideal biochemical analyser which could measure the concentration of any relevant substance at any time and at any location would be limited to measuring changes in the local concentration which can then be correlated with developmental events. We would still have no insight into why these changes have occurred, what the driving force is, what the cause is, and what the effect is—these questions would remain in the dark.

Most of the information about how development is controlled has been derived from experimental interference with developing systems; for instance by removal or irradiation of tissue or by transplantation of pieces into unnatural positions. The lack of knowledge about the basic mechanisms by which development is controlled is especially surprising in view of the large amount of detailed data about how a particular organism will react after a particular experimental manipulation.

As in any other branch of science, attempts have been made to explain the observed details through the invention of hypothetical mechanisms—models—which account for the observation as well as possible. A criterion of a good model is that several seemingly unrelated observations appear as the expression of a single underlying mechanism. If the assumed mechanism is simple, fewer parameters are involved. There is then less danger that a wrong mechanism will fit the observations by the choice of convenient values of the parameters. Furthermore, it is desired that the assumed mechanism is molecularly feasible and compatible with the known laws of physics and chemistry.

To compare the experimental results to a hypothetical model, it is invaluably helpful to have the model in a precise mathematical form. Only then discrepancies between the intuitively reasonable and the actual properties of the model, between the model and the experimental results, can be detected. Mathematically formulated models, which are usually called theories, have been a common tool in physics for a long time but still lack an adequate place in developmental biology. The superior power of a mathematically formulated model versus a verbally formulated model may be recalled by citing a well-known example from physics. The orbits of the inner planets were so precisely described by Newton's theory that deviations found in the orbits of outer planets compelled the assumption of additional planets—which were subsequently observed. It is evident that such a prediction would never have been possible by a non-mathematical model or by precise observations alone. To give another example, the Hydrogen atom emits light at only certain defined wavelengths. It would be impossible to determine from this observation alone that the electrons are restricted to certain discrete orbits around the nucleus. It was only the precise fit of the observed spectral lines with Bohr's model that made this hypothesis acceptable despite its variance with classical physics.

The attempt to describe the control of such a complex phenomenon as development of a higher organism with a limited number of interacting substances and, therewith, with a limited number of mathematical equations may at a first glance appear unrealistic. The situation becomes more promising if a separation into easier-to-understand submechanisms is possible. Such mechanisms are gene regulation, cell shape changes, cell movement, cell recognition by surface properties, etc. A further separation is possible with respect to different parts of an organism. For instance, the development of an arm and a leg can be regarded as independent processes. Even the antero-posterior axis and the proximo-distal axis of the same vertebrate limb can be dealt with separately. However, this separation is only an approximation and a closer look usually reveals their mutual dependence. This is not surprising, since the many parts have to come together to form a harmonious entity. Therefore, one should not be surprised if a simple model, explaining precisely many features of a particular developmental system, is only an approximation. The necessary complications do not argue against the model, but should be partially expected for the integration of the individual aspects into unified systems.

In a developmental system a signalling and signal-receiving mechanism must exist which enables the cell to develop in a manner appropriate to its position. The necessity of such a signal system is easily demonstrated by the fact of regeneration. The removal of a part of an organism can elicit a signal in the remaining parts which initiates a development through a completely different pathway than normally would be followed. This, of course, must require some kind of spatial communication. The goal of this book is to show which interactions of substances can lead to such signalling systems and how the cells can then respond to these signals in order for stable states of determination to be attained. The behaviour of the proposed reactions will be compared with experimental observations to demonstrate that many different observations are explainable under a few assumptions.

2

Some basic features of control of development

Intensive research in developmental biology has yielded evidence for some general control mechanisms, such as: gradients (Boveri, 1901; Morgan, 1904; Child, 1929), morphallaxis and epimorphosis (Morgan, 1901), the organizer (Spemann, 1938; Child, 1946), the embryonic field (Weiss, 1939), inhibitory fields (Morgan, 1904; Schoute, 1913), embryonic induction (Spemann, 1938; Saxen and Toivonen, 1962) and positional information (Wolpert, 1969, 1971) all of which overlap one another to a greater or lesser degree. The experimental facts from which these principles are deduced have been repeatedly reviewed in recent years (Sondhi, 1963; Wolpert, 1969, 1971; Robertson and Cohen, 1971; Cooke, 1975b). There are also excellent reviews available for particular developmental systems, such as insects (Counce and Waddington, 1972; Sander, 1976; Lawrence, 1970), especially *Drosophila* (Ashburner and Wright, 1978), the sea urchin (Hörstadius, 1973; Czihak, 1975), planarians (Chandebois, 1976a), hydroids (Webster, 1971; Tardent and Tardent, 1980) and the vertebrate limb (Hinchliffe and Johnson, 1980).

To have a basis for the discussion of the mechanism which could be responsible for the control of development, some of these principles are briefly discussed together with some key experiments from which they have been derived.

Spatial differences (pattern) must be generated during development

Although the eggs of higher organisms have some structures, the final complexity of an organism cannot be contained in the egg in some hidden form. A separation of the two blastomeres after first cleavage of a sea urchin egg can lead to two complete and well-proportioned embryos. These eggs are

4

therefore clearly not merely a mosaic, and the final structure must arise by an internal, self-regulatory process. Similarly, any form of regeneration requires that the intricate spatial arrangement of the regenerate is formed from less structured tissues. Therefore, any mechanism assumed to explain development must be able to generate spatial differences from more or less homogeneous conditions.

Organizing centres and their induction

The developmental fate of a larger region is frequently controlled by a small group of cells. The implantation of a small piece of the dorsal lip of the amphibian blastopore into the ventral site of the blastula can induce the formation of a second embryo. The discovery of the amphibian organizer by Spemann and Mangold (1924) led to a great optimism that substances controlling development may be isolated. It was a very surprising and disappointing outcome that very unspecific stimuli such as the implantation of foreign or denatured tissue can also lead to an induction. Another example of an unspecific induction is the formation of a second abdomen. After puncturing or UV-irradiation of the anterior side of an egg of the midge *Smittia*, a second abdomen is formed there instead of a head (see Fig. 8.4). Due to leakage or radiation damage, these manipulations can only lower but not increase the concentration of a particular substance. Most interestingly, in this case, the segment pattern is changed in more than one half of the egg although only less than a quarter of the egg was subjected to the experimental interference, indicating the long range of determinative influence on the surrounding tissue.

Induction of a new organizing region is frequently an all-or-nothing event. If successful, the final outcome is almost independent of the stimulus. In hydra a small piece of near-head tissue can induce a new head. Shift of posterior pole material in the egg of a leafhopper (see Fig. 8.2) can induce the complete sequence of structures and especially an abdomen at unusual location. After some time, the inducing material can even be ligated off and the formation of structures proceeds further. Something has spread out from the incipient organizer and "infected" the surrounding tissue, indicating that the formation of a new organizer, once initiated, is both autonomous and self-regulating.

The term "induction" is used for at least two phenomena which have different properties and certainly a different molecular basis and should not be confused. On the one hand, it signifies the triggering of a new organizing region, as mentioned above. On the other, it denotes the formation of a new structure caused by a direct contact of at least two different cell types, the induction of a lens in the ectoderm by the outgrowing eye cup, for instance.

Inhibitory fields

The formation of a structure signifies that something is formed at a particular location which is not formed in the environment. Botanists had realized very early that each leaf primordium is surrounded by a zone where the formation of further leaves is inhibited (Schoute, 1913). The spacing of bristles of insects has also been explained in a similar way (Wigglesworth, 1940). Similarly, in hydra a new head (or more precisely, a new hypostome) can be induced by the implantation of near-head tissue into the body column. The probability of inducing a new head decreases with decreasing distance from the existing head. This forces the conclusion that an inhibitory effect emanates from some existing structures and that this inhibition diminishes with increasing distance from its origin. On the other hand, not every structure is surrounded by an inhibitory field. In many developmental systems, mirror-symmetrical duplications are formed. An example is the formation of a double abdomen just mentioned. At the line of symmetry, identical structures are formed close to each other, obviously without a mutual inhibition.

Polarity

Most tissues have a polarity, an asymmetric distribution of a particular property over a field of cells. Polarity can have at least three different origins: (1) the cell composition changes over the field; e.g. in a hydra, the nerve cell density is much higher in the head region than in the body column. (2) A homogeneous appearing field may in fact be subdivided into a sequence of distinguishable structures. This seems to be the case in the abdominal or in the leg segments of an insect. (3) The individual cells are polar and aligned. The basis of an observed polarity can frequently be experimentally determined. For instance, a leg segment of an insect consists of a sequence of structures, since missing structures are replaced by intercalary regeneration. In addition, the individual cells have a proximo-distal polarity indicated by the orientation of bristles. Transplantation experiments (see Fig. 13.1) indicate that the sequence of structures determines the orientation of the individual cells, not the other way round. Similarly, dissociation and reaggregation experiments with hydra indicate that the cell composition, and not the orientation of individual cells is the polarity-determining factor (Fig. 6.4). Polarity leads to a predictable behaviour of tissue fragments. For instance, head regeneration in small fragments of planarians appears at the site oriented originally towards the head. Similarly, heads of hydra or hydrants of tubularia always appear at the distal end. Morgan (1904) concluded from experiments with tubularia that the decision to form a hydrant at a particular location results from a mutual competition of all the cells of the tissue to form that structure. He interpreted polarity as a graded advantage in this competition.

Prepattern, gradients, morphogenetic fields, positional information and sequences of structures

If a new structure emerges during development, this may be caused directly by a communication between cells or could result from a two-step process in which a localized high concentration of a "morphogen" is first formed, which in turn initiates the determination and differentiation at that particular area. In the latter case, the morphogen distribution may be considered as a pre- or primary pattern which precedes the structure to be formed.

Evidence for the existence of a prepattern has been derived from rather different organisms. With genetic mosaics, Stern (1956) has shown for bristle-formation in *Drosophila* that mutants exist in which the response of the cells but not the formation of the prepattern is altered, indicating clearly the separation of the two processes. In hydra, after removal of a head, a new head appears at about 48 hours. However, important changes take place in the future head region within the first four hours after head removal. Transplantations of small pieces of tissue from the prospective head region into the body column of another animal lead to different results depending on whether the transplantation is performed immediately or after about four hours. Only in the latter case, a new head will be induced (Webster and Wolpert, 1966). The four-hour time interval is probably too short to allow much cell differentiation but is long enough for the local production of a substance which initiates head formation even after transplantation into a new environment.

A prepattern can determine more than one structure if the morphogen is not used for an all-or-nothing decision. A graded distribution of a morphogenetic substance can specify a sequence of structures as a function of the local morphogen concentration. Wolpert (1969, 1971) has generalized this idea in the concept of positional information. Such a mechanism has been suggested by investigations on the antero-posterior axis of insects (Counce and Waddington, 1972; Sander, 1976), the dorso-ventral axis of insects (Nüsslein-Volhard, 1979; Nüsslein-Volhard et al., 1980), the organiz-ation of the antero-posterior axis of the amphibian limb (Slack, 1977a,b) and of the chicken wing bud (Tickle et al., 1975). These systems suggest that a long-ranging signal exists which spreads out from a small group of cells—the organizing region.

In other cases, the spatial arrangement seems not to be controlled by a two-step process: generation of positional information and its interpretation but results from a direct and short-ranging communication of adjacent cells controlling in this way the correct neighbourhoods. The organization of an individual insect segment (see Fig. 13.1) seems to be of this type.

Attempts toward a theory of development

Development must be ultimately a biochemical process, consisting of interactions and movement of molecules. Turing (1952) made the very important discovery that spatial concentration patterns can be formed if two substances with different diffusion rates react with each other. This is entirely contrary to our intuition since with diffusion one associates a smoothing of any local accumulation of molecules but not the creation of such concentration maxima. To make analytical solutions possible, Turing had linearized his equations, which led to some problems with unlimited growth of the pattern. In respect to non-linear equations he wrote (Turing, 1952, p. 72):

> The difficulties are, however, such one cannot hope to have any very embracing theory of such process, beyond the statement of the equations. It might be possible, however, to treat a few particular cases in detail with the aid of a digital computer. This method has the advantage that it is not so necessary to make simplifying assumption as it is when doing a more theoretical type of analysis.

Numerical solutions of Turing's non-linear equations show periodic distributions that are stable in time (Martinez, 1972). Meanwhile, many attempts have been made to analyse mathematically such reaction-diffusion mechanisms (see, for instance, Gmitro and Scriven, 1966; Kopell and Howard, 1973; Bard and Lander, 1974; Babloyanz and Hiernaux, 1975; Granero et al., 1977; Fife, 1979; Lacalli and Harrison, 1979). The connection of this type of reaction with irreversible thermodynamics has been worked out by Prigogine and Nicolis (1971). The importance of the amplification of small deviation from an average for all sorts of pattern formation (also non-biological) has been pointed out by Maruyama (1963). A basically different mechanism for the generation of sequences of structures has been proposed by Goodwin and Cohen (1969) based on the phase difference of two waves, propagating with different velocities.

By considering known principles of biological development—the fields of inhibition and the possibility of local induction—we have shown that short range autocatalysis coupled with long range inhibition gives rise to spatial concentration pattern with features known from biological systems (Gierer and Meinhardt, 1972, 1974; Meinhardt and Gierer, 1974; Gierer, 1977a,b; Meinhardt, 1978a). Such interactions can be given a mathematically precise form of coupled differential equations which allow molecular interpretations.

This principle allows us to predict even for highly non-linear interaction whether or not a stable pattern can be formed. The properties of these systems will be illustrated by comparing their behaviour (numerical solutions of the differential equations) with a large variety of biological observations. We will examine these models further in the following sections.

3

Self-enhancement (autocatalysis) and long range inhibition—a general mechanism of pattern formation

One of the most fascinating aspects of development is the generation of structures from a more or less homogeneous egg. This was felt to be so miraculous that a long argument arose as to whether the laws of physics were sufficient for an explanation of development. Driesch, for example (see Herbst, 1942)—who made such great discoveries, e.g. that the cell nuclei in a developing sea urchin egg are totipotent and therefore the interaction of cytoplasm with the nuclei is decisive for the developmental pathway— believed in a vital power which was assumed to be unexplainable by physical laws.

However, a look at inorganic nature reveals that formation of pattern is not peculiar to living objects. Pattern formation is the rule also in the non-living world. Formation of galaxies, stars, clouds, rain drops, lightning, river systems, mountains, crystals, all forms of erosion—all testify to the generation of ordered structures (Fig. 3.1). It is instructive to look for common principles in the generation of these structures. If a small deviation from a homogeneous distribution has a strong positive feedback on itself, the deviation will increase. For example, erosion proceeds faster at the location of some random initial injury and sharp contoured rivers are formed despite the rain being distributed almost homogeneously over the country. A large sand dune may result from a stone in a desert which produces a wind shelter and may thus locally accelerate the deposition of sand; this deposition increases the probability of further sand deposition, and so on.

As well as the strong positive feedback—autocatalysis—another element is required for pattern formation: lateral inhibition. Once an autocatalytic centre has arisen, a suppression of the autocatalysis in the neighbourhood must occur, otherwise the reaction would spread like a grass fire. In a grass fire, all of the available fuel in an area would be converted to ashes, leaving no pattern to the landscape. The strong short-range positive feedback must therefore be supplemented by a longer-range negative feedback. In inorganic pattern formation the antagonistic reaction to the self-enhancement is in most cases based on the using-up of a movable prerequisite. Examples are given in Fig. 3.1. Pattern formation requires energy. The wind has to blow to

Fig. 3.1. Pattern formation in non-living nature, based on self-enhancing (auto-catalytic) processes. (a) *Sand dunes* may arise from a sand deposit behind a small wind shelter; this deposit increases the wind shelter and thereby accelerates further deposition. (b) *Erosion* proceeds faster at some injury. More water collects in the incipient valleys, accelerating the erosion there. (c) *Thunderstorm clouds*: warmed up air expands, becomes lighter and moves upwards. There, the surrounding air is even colder and this accelerates the upsteam. (d) *Waves* in layers of down-streaming water are formed despite a uniform rainfall. The speed of the water depends on friction with the ground. A thicker layer has less friction and is therefore faster. Any local increase leads to a local acceleration and the water then catches up with water already further down, amplifying the local increase and the speed further. (e) *Lightning* with sharp contours are formed despite the fact that charged clouds are diffuse. Under the influence of the voltage-difference ions are accelerated, producing more ions by collision with other atoms, leading to an avalanche effect. In all these situations, an effect of lateral inhibition is involved too. If the sand or water accumulates at a particular location, it is depleted at others. The upstreams at some location necessitates down streams at others.

form a sand dune; water has to be evaporated to form clouds, raindrops and rivers.

An indication that the same principle—local autocatalysis and lateral inhibition—is involved in biological pattern formation may be seen in the appearance of additional heads in hydra. As mentioned, a small piece of tissue which was derived from near the head and grafted into the body column can form (together with some adjacent tissue) a second head. The development of the complete new head was initiated by a small disturbance. After a while, the implanted tissue can even be removed, and the induced head formation proceeds autonomously. Something has "infected" the surrounding tissue which is then able to form the new head on its own. If this infection takes place— why is not the whole hydra infected? The existing head has an inhibitory influence on the induction of a secondary head which limits the spreading of the infection. The inhibition is highest in the head area but is detectable throughout the animal. We will discuss this system in more detail on p. 48.

Activator and inhibitor

To apply this principle to biochemical reactions, let us assume a substance—to be called activator a—which stimulates its own production (autocatalysis) and the production of its antagonist—the inhibitor h. To carry out the necessary long-range inhibition, the inhibitor must diffuse more rapidly. In an extended field of cells a homogeneous distribution of these substances is unstable, since any small local elevation of the activator concentration—resulting perhaps from random fluctuations—will be amplified by the activator autocatalysis. The inhibitor, which is produced in response to the increasing activator production, cannot halt the local increased activator production, since it diffuses quickly into the surrounding tissue and suppresses activator production outside the activated centre. Thus, the locally increased activator concentration will increase further, with increasing concentration, the maximum becomes narrower and narrower until some limiting factor comes into play, for instance, the loss of activator from the narrow peak by diffusion becomes equal to the net production. A stable activator and inhibitor profile is ultimately obtained, although both the substances continue to be made, to diffuse, and to be broken down. Such a simple system of two interacting substances is, therefore, able to produce a stable, strongly patterned distribution from a nearly homogeneous initial distribution, as it occurs in biological pattern formation. As demonstrated below, the resulting pattern can be monotonic or periodically varying over a region depending on the parameters such as diffusion and lifetime of the substances involved. The extension of an area with a high substance concentration can be proportional to the total size. The pattern can be stable or oscillating in time, the location of a high concentration can be directed by minor internal or external asymmetries. An active centre which has been removed can "regenerate". A second centre can be induced by rather unspecific manipulations. An activator centre therefore has the essential properties of a classical organizer.

Interactions which lead to stable pattern

Development is alteration in time. Assuming that development is controlled by substances, any theory of development has to describe concentration changes of substances as a function of other substances involved and as a function of spatial coordinates and time.

Two conditions have to be fulfilled before a stable pattern can be generated. (1) A local deviation from an average concentration should increase further, otherwise no pattern would be formed, and (2) the increase should not go to infinity. Instead, the emerging pattern should reach a stable steady state. This is possible if an increase in one part is necessarily connected with a decrease in another part of the field. The effect of this is that the total amount of substances in the field are roughly conserved.

Let us think of the simplest possible interactions which can lead to a stable pattern. The reader not familiar with mathematical expressions should not be worried about the following differential equations. An equation is an unambiguous expression of what the hypothesis is. We will not discuss any mathematical problems but use the equations as a shorthand of the assumed interactions. In most of the equations, a change per time unit is described. For instance, $\partial a/\partial t$ denotes the change of a per time unit. The overall time course of a concentration is calculated by integration of such a differential equation. In other words, the concentration change after a short time interval is added to a given initial condition, leading to a new concentration profile, determining the following change, which is added again, and so on.

The simplest assumption one can make about the disappearance of the molecules is that the number of decaying or escaping molecules is proportional to the number of molecules present ($\partial a/\partial t = -\mu a$) similar as to the number of individuals dying per day in a city depends on the number of people living there. Concerning the communication of a cell with its neighbours, a simple assumption is diffusion. Let us assume a cell i with two neighbours, $i-1$ and $i+1$. The corresponding concentrations are a_i etc. The net exchange between two cells depends on the concentration difference, e.g. $a_{i-1} - a_i$. The gain or the loss of the cell i from or to its left and right neighbours is therefore $D_a((a_{i-1} - a_i) + (a_{i+1} - a_i))$ where D_a is the diffusion constant. In a linear concentration gradient, the net exchange is zero since a

cell loses by diffusion the same amount to the lower neighbour as it gains
from the higher neighbour. If the space is continuous and a subdivision into
individual cells is impossible, diffusion is proportional to the second
derivative $(\partial^2 a/\partial x^2)$.

If such simple modes of decay and redistributions are assumed, the critical
terms for the pattern-forming reactions are the production terms as function
of the other substances involved. We have derived a general criterion about
which interactions lead to a stable pattern and which do not (Gierer and
Meinhardt, 1972). Assuming an activator a and an inhibitor h, the following
example satisfies this condition:

$$\frac{\partial a}{\partial t} = \frac{ca^2}{h} - a\mu + D_a \frac{\partial^2 a}{\partial x^2} \qquad (3.1a)$$

$$\frac{\partial h}{dt} = ca^2 - vh + D_h \frac{\partial^2 h}{\partial x^2} \qquad (3.1b)$$

We can easily gain some intuition about why this reaction leads to a pattern.
First let us look at eq. 3.1a and assume that the inhibitor concentration is
constant (and arbitrarily equal to 1) and that activator distribution is uniform
(no change by diffusion). Setting the constants c, μ and v arbitrarily to unity,
we get

$$\frac{\partial a}{\partial t} = a^2 - a.$$

We get a steady state $(\partial a/\partial t = 0)$ at $a = 1$. However, this steady state is
unstable. For example, if a is slightly larger than 1, $\partial a/\partial t$ is positive, that
means a will increase further. To achieve this instability, it is obvious that the
production term has to be of higher order than the linear decay term, that
means non-linear. By itself, such autocatalysis would lead to an overall
explosion.

Now let us include eq. 3.1b. The inhibitor production is controlled solely
by the activator concentration present in the system. Let us assume a very
rapid equilibrium of the inhibitor to a given activator concentration. At the
steady state $(\partial h/\partial t = 0)$ the inhibitor concentration will be $h = a^2$. Inserting
this into eq. 3.1a we get

$$\frac{\partial a}{\partial t} = \frac{a^2}{a^2} - a = 1 - a$$

and again a steady state at $a = 1$. This steady state is stable since, for
instance, if a is larger than 1, $\partial a/\partial t$ is negative and therefore the deviation
from the equilibrium will be regulated back to $a = 1$.

So far we have two extreme cases, either stability or instability, depending on whether we include the action of the inhibitor or not. By a convenient choice of diffusion rates we can achieve local instability with overall stability of the system. As discussed in the last section, the autocatalysis must be a more or less local process while the inhibition must spread out rapidly. If we now follow a local activator increase, this leads according to eq. 3.1 to a locally increased inhibitor production. However, due to the high diffusion rate, the inhibitor surplus becomes widely distributed. Therefore, despite the local activator increase, the inhibitor concentration can be regarded as constant and this leads, as we have seen, to a further increase of the activator: The pattern begins to form (Fig. 3.2b).

With the further local activator increase, the inhibitor concentration can no longer be considered as constant. Its overall increase leads to a decrease of the activator production outside of the activated area. Also at the activated site, the activator increase is restricted, since, as the developing maximum becomes steeper and steeper, the loss by diffusion increases and the surrounding "cloud" of inhibitor finally stabilizes the pattern.

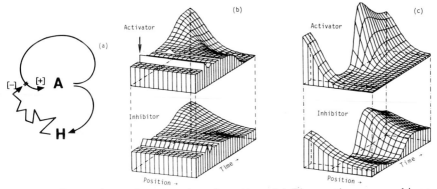

Fig. 3.2. Generation and regeneration of a pattern. (a) The reaction: assumed is an autocatalytic substance—the activator (A)—and its more rapidly diffusing antagonist—the inhibitor (H). (b, c) Pattern formation in a linear array of cells. In this and most of the following figures, the activator (top) and inhibitor concentration (below) in a linear array of cells is plotted as function of position and time. (b) A completely uniform distribution of both substances is stable, an artificial activator increase (arrow) disappears following a compensating increase in the inhibitor concentration. But a small *local* activator increase or even a random fluctuation cannot be compensated, since the additionally produced inhibitor diffuses quickly into the surroundings. In the slightly activated region, the activator concentration increases further until a steady state is reached in which the gain by production and the loss by diffusion and decay is balanced. (c) Such a pattern has strong self-regulatory properties, e.g. it can "regenerate". With the removal of the activator maximum, the remnant inhibitor decays and a new maximum is formed via autocatalysis. Depending on how much of the former activator maximum is included in the fragment, the polarity can be maintained (c) or reversed (see Figs 4.4 and 4.7).

This local high concentration can be used as a signalling system, for instance, to initiate head formation in hydra. A pattern formed in this way has strong self-regulatory properties. For instance, after removal of the activated area, the remaining inhibitor decays and the activator maximum "regenerates" via autocatalysis (Fig. 3.2c).

For many calculations shown in this book, eq. 3.1 is used with a few additions which play an important role in the theoretical description of real biological systems

$$\frac{\partial a}{\partial t} = \rho \frac{a^2}{(1 + ka^2)h} - \mu a + D_a \frac{\partial^2 a}{\partial x^2} + \rho\rho_0 \tag{3.2a}$$

$$\frac{\partial h}{\partial t} = \rho'a^2 - vh + D_h \frac{\partial^2 h}{\partial x^2} + \rho_1. \tag{3.2b}$$

The small basic (activator-independent) activator production ρ_0 can initiate the autocatalysis in areas of low activator concentration. As we will see, this term is important if new centres are to arise during growth or regeneration. In contrast, the basic inhibitor production ρ_1 can suppress the appearance of secondary maxima, a feature which is important if an ordered sequence of structures is to be specified by positional information (p. 62). The factor ρ denotes the source density. Its polarity-determining influence will be discussed on p. 37. If the activator production saturates at a high concentration due to a term $1/(1 + ka^2)$, the activated area is size-regulated (p. 39).

Equation 3.2 is of course only an example. The general condition of local instability *and* stable average concentration provides an easy check whether a particular reaction will lead to a stable pattern or not. For instance, a linear stimulation of the inhibitor production by the activator but a non-linear action of the inhibitor leads to a stable pattern as well (eq. 3.3):

$$\frac{\partial a}{\partial t} = \rho \frac{a^2}{h^2} - \mu a + D_a \frac{\partial^2 a}{\partial x^2} \tag{3.3a}$$

$$\frac{\partial h}{\partial t} = \rho'a - vh + D_h \frac{\partial^2 h}{\partial x^2}. \tag{3.3b}$$

This interaction allows an interesting molecular interpretation: the activator molecule decays or is converted into an inactive molecule which acts as inhibitor, since it is in competition with the undegraded activator molecules.

Other molecular realizations of the principle of autocatalysis and lateral inhibition are possible. For instance, the inhibitory effect may be realized by a depletion of a substrate or of a precursor which is consumed in the autocatalysis. This reaction has some properties different from the reaction scheme eq. 3.2 which could allow a distinction on the basis of experimental observations (see p. 34). The autocatalysis may be realized by the mutual inhibition of two substances (p. 36). Further, the autocatalysis can be a reaction chain consisting of many elements or can be realized by an autocatalytic release of bound substances rather than by a control of the production. To achieve necessary non-linearity of eq. 3.2, two activator molecules can form a dimer or can cause an allosteric shift which releases a further activator molecule. A single inhibitor molecule would block such release. The redistribution of the inhibiting substance by diffusion may be enhanced by convection.

4

Polar, symmetric and periodic patterns— basic properties of an activator–inhibitor system

Generation of polar and symmetrical structures

The mechanism of pattern formation requires that the inhibitor diffuses more rapidly than does the activator. Let us first assume a linear array of cells. In an area small compared to the range of the activator (i.e., the mean distance between the production and decay of the activator molecule), the different diffusion rates cannot come into play since both substances equilibrate rapidly in the small field. Therefore, in a small field, only a homogeneous distribution at a well-defined concentration is possible.

In the growing field or in a field in which the diffusion rates decrease, for instance, by a progressive cleavage, the homogeneous distribution becomes unstable quite abruptly and a stable graded concentration pattern emerges. For biological applications it is important that at first only a marginal (asymmetric) activator maximum is possible or, in other words, that graded distributions are formed (Fig. 4.1a). This is due to the fact that a marginal activator maximum requires space for only one activator slope, while a central maximum would require space for two slopes which is not available when the critical size is just reached.

The orientation of the emerging pattern is usually determined by internal or external asymmetries. The eggs of most species mature in an asymmetric environment and become in themselves non-homogeneous, which directs the pattern in a predictable way and no symmetry breaking is required. A rare exception is the almost homogeneous egg of the brown alga *Fucus* (Fig. 4.2). The development of the spherical egg begins with the outgrowth of a so-called rhizoid at one particular side. The outgrowth can be directed by light, electric current, differences in pH or temperature between the different sides

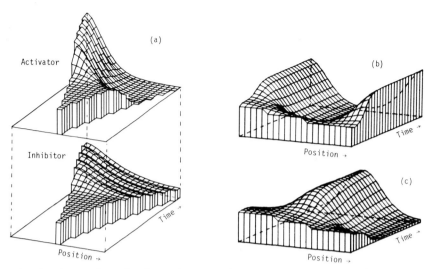

Fig. 4.1. Formation of polar and symmetric pattern. (a) A monotonically graded distribution of the activator and inhibitor is formed in a growing field of linearly arranged cells. A pattern can only be formed when a critical size (range of activator) is exceeded. At the critical size the maximum can appear only at one boundary and can remain there upon further growth. Therefore, the model explains the formation of polar concentration profiles, even from an initially near-homogeneous situation. (b, c) At larger extensions of the field, the formation of symmetrical patterns is favoured even if the initial distributions are clearly asymmetric. Minor differences are decisive whether a maximum appears at the centre or at both margins. Experimental interference during development leads frequently to an abnormal symmetric pattern, e.g. in *Fucus* (Fig. 4.2), in sea urchins (Fig. 4.3), in stolons of hydroids (Fig. 4.7) and in insects (Fig. 8.4).

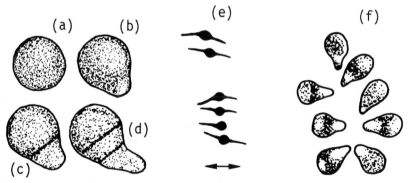

Fig. 4.2. Outgrowth of the brown alga *Fucus*. As a rare case, the egg of the brown alga *Fucus* has almost no internal asymmetry. The normal outgrowth (a–d) can be oriented by differences of temperature, pH, light or electric potential. (e) Illumination of an egg with polarized light (the arrows indicate the plane of vibration of the electric vector) can lead to a symmetric outgrowth. (f) The influence from other eggs is sufficient to orient the outgrowth. After Jaffe (1968).

of the egg, or by mutual attraction from other nearby eggs (Jaffe, 1968). This supports the view that there is an unstable situation and that any asymmetry can orient the pattern formation. In the absence of any orienting effect, the outgrowth appears at random but much delayed. This is also in agreement with the proposed mechanism, since the time required for the formation of an activator peak is shorter for larger deviations from the semistable equilibrium.

In a field of the size comparable to the activator range, the activator maximum depends on the total size of the field, whereas in larger fields, it becomes only a small fraction of the total field and is almost size-independent. Therefore, it could be favourable for biological systems to utilize fields much larger than the activator range. Organizing centres with properties of an activator maximum, for instance, the amphibian organizer or the activation centre in insects (see Fig. 8.1) occupy indeed only a small part of the field.

To maintain a simple polar pattern during further growth or shrinkage of the diffusion range, it is required that (1) the maximum remains at the margin and (2) that no secondary maxima are formed. The first requirement can be satisfied in an activator-inhibitor system, but creates problems in pattern forming systems which are based on the depletion of a precursor (Fig. 5.1). The appearance of secondary maxima can be prevented by a baseline inhibitor production (ρ_1; eq. 3.2) which enables a second steady state at low activator concentration. This low steady state is stable as long as a critical activator concentration is not surpassed. Such a safeguard mechanism against secondary maxima has its price. The basic inhibitor level can suppress the "regeneration" of an activator maximum. Perhaps the lack of "regeneration" after complete removal of the "activation centre" at the posterior pole of an insect egg (see Fig. 8.1) has this origin. The baseline inhibitor level offers the possibility that the gradient-forming mechanism is "asleep" until a particular developmental stage is reached independent of the extension of the field; an example will be given in Fig. 4.8b.

If a gradient system is disturbed when a larger size is already obtained, a symmetric distribution is favoured. The high activation appears either at the centre or at both margins. After such a disturbance, the distributions may still be quite asymmetric but the self-regulatory features lead to one of the two symmetric forms (or re-establish the monotonic distribution). Minor differences can decide which of the few possible patterns will develop. Symmetric development has been observed after experimental disorganization, e.g. after centrifugation of insect eggs a development of symmetric embryos with either two abdomina or two head lobes occur, the latter with surprisingly little further segmentation (Yajima, 1960; Rau and Kalthoff, 1980).

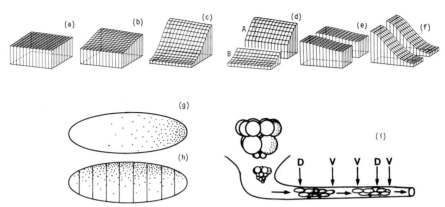

Fig. 4.3. Formation of a graded distribution in a two-dimensional field. (a–c) When the size of the field is of the order of the activator range, the maximum appears at one boundary (similar to that in Figs 3.2 and 4.1). Minor asymmetries can orient the gradient. The alignment occurs preferentially along the longest extension of the field. A sequential pattern formation and interpretation (p. 108) can generate orthogonal coordinate systems. For instance, under the influence of the gradient, the field is subdivided into narrower strips (A, B in d), the pattern reorients itself perpendicular to the original orientation (e, f) despite the initial asymmetry within the two parts. Autocatalysis and lateral inhibition is therefore convenient to organize the long antero-posterior axis of an insect egg (g). For the shorter dorso-ventral axis (h), an initial subdivision into stripe-like "segments" would be required. (i) A biological example for the orientation of a pattern by the change of the geometrical form is the orientation of the dorso-ventral (D-V) axis of a sea urchin embryo by squeezing it through a capillary. The front side of the sausage-like deformed embryo usually becomes the ventral pole. In some cases, symmetrical embryos are formed (Lindahl, 1932).

In a two-dimensional field of a size comparable to the activator range, a graded distribution in one dimension and a constant distribution in the other dimension can be formed (Fig. 4.3). As a rule, the gradient will be oriented in the direction of the largest extension of the field. The mechanism offers therefore a possibility to detect the longest dimension of a field and to organize it. A sequential formation of gradients enables the formation of orthogonal coordinate systems (Fig. 4.3). An alternative mechanism which allows gradient formation along a small axis of a field will be discussed later. It depends on lateral activation of two substances (Fig. 12.4).

Regeneration and induction of activator maxima

The local high concentration of the activator or inhibitor can serve as a signal to form a particular structure. It can, for instance, initiate cell differentiation or tissue evagination. Characteristic of many embryonic systems is their ability to regenerate parts after removal. One possibility is a reprogramming of the remaining tissue (morphallaxis) for which the model provides an explanation. With the removal of the activated site, also the site of inhibitor production is removed, the remaining inhibitor decays and via autocatalysis a new maximum is formed (see Fig. 3.2c). The autocatalysis requires that at least some activator is present, originating either from the remaining activator or from a small leakage-like baseline activator production (ρ_0 in eq. 3.2). The self-regulating activator-inhibitor system is therefore able to ensure that at least one "organizing region" is present in the field. The regeneration of a hydra head (Fig. 6.2) or the ventral side of a sea urchin (Fig. 4.4) will be discussed below in more detail.

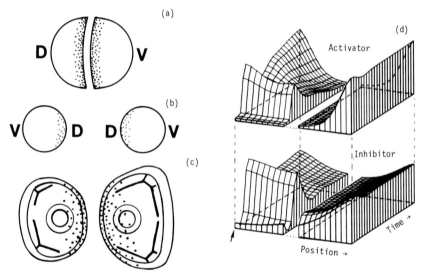

Fig. 4.4. Reversal of the dorso-ventral polarity in a fragment of a sea urchin embryo. (a–c) Experimental evidence: separation of the dorsal and ventral half of a sea urchin embryo in the 16-cell or in the early blastula stage; the cells close to the area of separation are mildly vitally stained (speckles in part a). In the pluteus larvae, both stained regions form a dorsal site, indicating a polarity reversal in one of the fragments. The more advanced development of the ventral side indicates that it is the original dorsal half which has reverted the polarity (see Hörstadius, 1973). (d) Explanation in terms of the model. The development of an activator-inhibitor distribution after a separation in an activated and in a non-activated fragment is shown. In the latter, the new activation is formed at the side with the lowest concentration of remaining inhibitor (arrow) and therefore with a reversed orientation.

If the size of the field is large enough to accommodate more than one activator slope, additional maxima can be induced either by a small local increase of activator, realized for instance by the implantation of activated tissue or unspecifically by the removal of inhibitor. The probability of inducing a secondary maximum with a given stimulus is expected to depend strongly on the distance between the site of manipulation and the natural organizer (activator maximum), since the local stimulation has to overcome the inhibition arising from the maximum. A removal of the existing maximum facilitates the induction of a new activator maximum considerably, since after such removal, the level of inhibition drops in the field. The possible induction of a new maximum is an all-or-nothing event, the final maximum being independent of the mode of stimulation. The concentration of activator increases via autocatalysis only if sufficient activator is present to overcome the inhibition from the existing maximum. This developing maximum will increase until it comes to a steady state with the self-produced inhibitor. In this process, tissue in the environment of the stimulated site can be "infected" to participate in the autocatalysis. Therefore, after a period of contact, the inducing tissue can be withdrawn and nevertheless the formation of a second maximum proceeds autonomously.

Unspecific induction

After the discovery by Spemann and Mangold (1924) that the dorsal lip of the amphibian blastula can induce a second embryo, much enthusiasm arose concerning isolation of substances which are responsible for the induction. As mentioned, it was surprising and disappointing to discover that very unspecific stimuli lead to an induction; even cell poisoning, injury, or implantation of killed or foreign tissues is sufficient. Waddington et al. (1936) proposed that this unspecificity results from the removal of an inhibitor. In the model, a local activator maximum is necessarily surrounded by a field of inhibition; the inhibition decreases with distance from the maximum. At higher distances, any artificial decrease can induce a new activation. Possible mechanisms for inhibitor decrease include leakage at an injury, breakdown due to a release of degrading enzymes, or destruction of inhibitor-producing structures by UV irradiation. The UV induction of the double abdomen malformation during early insect development (Fig. 8.4) will be given later as an example. Unspecific induction indicates that in these systems the inhibitory effect is derived from a real inhibitor and not from a depletion of a precursor (see p. 34). Unspecifically one can only destroy and not create molecules. Only the decrease of an inhibitor, not of a precursor or an activator can trigger the formation of a new maximum.

An example: the dorso-ventral organization of the sea urchin embryo

The dorso-ventral axis of a sea urchin embryo shows many properties of an activator-inhibitor system. While the animal-vegetal (p. 59) axis of a sea urchin egg is fixed during oogenesis, the orientation of the dorso-ventral (D-V) axis can be reoriented by mild external influences and the D-V pattern can regulate (for experimental details and literature, see Hörstadius, 1973; Czihak, 1975). After the first two cleavages of the developing egg, the four blastomeres are arranged like slices of an orange, each containing material from the animal as well as from the vegetal pole. Isolated blastomeres of a 4-cell stage embryo develop into small but complete larvae (Driesch, 1900) indicating that in each blastomere a new D-V axis is established (and that size regulation takes place). The reversed experiment is also possible. Two eggs pressed together can develop into only one very large embryo (Driesch, 1899; Boveri, 1901). The D-V axis can be reoriented by external manipulations. Local application of metabolism-inhibiting substances causes the formation of the dorsalmost area at this location, indicating that, in terms of the proposed model, it is the ventral side which corresponds to the activated area. After stretching of the cleaving egg, the D-V axis is oriented parallel to the artificial long axis (Lindahl, 1932). We have seen already in Fig. 4.3 the strong tendency of an activator-inhibitor pattern to orient itself according to the long extension of a field of cells. Minor asymmetries are decisive as to which end becomes the ventral side. For instance, the stretching can be accomplished by pressing the egg through a fine pipette, deforming it in a sausage-like manner. Usually it is the leading end which forms the ventral side (Fig. 4.3). If this end is poisoned by extensive staining with Nile Blue, the trailing end will form the ventral side.

After separation of a 16-cell embryo into a ventral and a dorsal half, the ventral half maintains its orientation, while in the dorsal half the D-V orientation is reversed. The originally dorsalmost area changes to the ventralmost area, while both the new dorsal areas are formed from material originally facing each other (Fig. 4.4). This is in good agreement with the proposed model, since the original dorsal area was exposed to the lowest inhibitor concentration. Therefore, after separation, the originally dorsalmost area becomes the newly activated (ventral) side. Condition for such a reversal is that the tissue is isotropic. As we will see later, in many tissues a very stable graded tissue property exists—the polarity proper—which orients a regenerating activator maximum according to this internal polarity (Figs 5.2, 6.2 and 6.3). Only in the absence of such a stable internal polarity is the transient gradient of the remaining inhibitor able to reorient the regenerating activator pattern.

The reversal of orientation in the dorsal half can take place also after a more extensive stretching but without a physical separation of both halves (Fig. 4.3i). In terms of the model, the distance between the activated (ventral) and non-activated (dorsal) site becomes so large that a second activation at the dorsal site can no longer be inhibited and a symmetrical embryo results. The reversal of the D-V axis in a dorsal half offers decisive evidence against the assumption that the overall orientation results from the alignment of individual polar elements arranged like the dipoles of a magnet (Driesch, 1900). In that physical analogy, each fragment of a magnet would retain its polarity.

Ranges of the activator and inhibitor in stolons of marine hydroids

Another biological system in which the basic features of the induction and maintenance of a structure can be easily compared with the expectations from the activator-inhibitor model are the colony-forming marine hydroids such as *Hydractinia* or *Eirene viridula*. The animals are interconnected by a branching network of hollow tubes, the so-called stolons (Fig. 4.5). A stolon elongates at a growing tip. A new branch can arise only at a distance of at least 400 μm from the tip. This suggests that the formation of a new tip is based on the formation of a new activator maximum and that the long-range inhibition emanating from the existing tip is responsible for the minimum spacing. If a growing tip touches an existing stolon, the mechanical stimulus induces a second tip at the existing stolon. After a while, the two stolons fuse and in this way an anastomosis is formed. A weak local pressure is therefore a natural stimulus for the initiation of a new tip. This pressure-sensitivity has been used by Plickert (1980) to measure the range of the fields. A new tip can be induced by a mechanical indentation at a distance of *c*. 230 μm behind the tip. If two stimuli are given very closely together (150 μm), then, in most cases only one tip is formed and it appears at a position *in between* the two stimuli. Thus, the location of the stimulus and the location of the outgrowth is not the same. This shows that the mechanical touching initiates a process which becomes independent of the stimulus. In the model, if two activator maxima are initiated too close to each other, the growing peaks would fuse at an intermediate position (Fig. 4.6a). The distance at which this averaging of stimuli is possible would correspond roughly to the range of the activator. At a somewhat larger distance of the two stimuli (200 μm), again only one tip is formed. But in this case, it appears at the location of one of the two stimuli. The other becomes suppressed. At even larger distances (280 μm), both stimuli are successful. This distance corresponds to the range of the inhibitor. The simulations (Fig. 4.6) show that behaviour is correctly described by the model.

Small fragments of a stolon regenerate. Three patterns are possible (Müller and Plickert, 1982): (1) the new tip arises at the side pointing towards the original tip; (2) the new tip is formed at the other end, leading to a reversal of orientation; and (3) both open ends form a new tip, leading to symmetrical development. The frequency at which these three possible types are formed changes drastically with the distance of the fragment from an original tip (Fig. 4.7); small distances lead to regenerates with normal orientation, larger

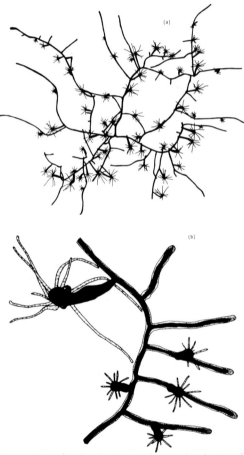

Fig. 4.5. (a) A colony of the marine hydroid *Eirene viridula*. The individual animals are interconnected by a network of stolons. (b) Details of a branching stolon with polypes (after Müller and Plickert, 1982).

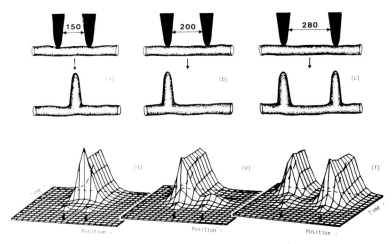

Fig. 4.6. Mechanical stimulation of stolon budding in the| marine hydroid *Eirene* (Plickert, 1980). (a) Depending on the distance (μm) between the two stimuli a single bud grows out either between the stimuli (a), at the location of one stimuli (b), or two buds are formed (c). (d–f) Explanation of the model. The stimulus is assumed to induce some local activator release (or inhibitor leakage). At small distances, the two emerging activator maxima fuse and form a central one (d). At larger distances, one maximum dominates and suppresses the other one (e). At even larger distances, both maxima can coexist (f).

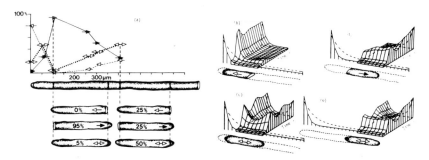

Fig. 4.7. Regeneration in reverted orientation in fragments of a stolon of a marine hydroid (*Eirene*) (Müller and Plickert, 1982). (a) Depending on the distance from an existing tip, a fragment of a stolon can either regenerate with the same polarity (◁─┤), in a reversed polarity (─▶) or in a symmetric way (◁─▷). |(b–e) Model: The normal tip is assumed to be controlled by an activator-inhibitor system. If some activator is included in the fragment (close to the tip), regeneration occurs according to the normal orientation (b). At larger distances, the graded inhibitor distribution leads to a reversal of orientation (d). In between these two areas, a narrow zone of transition exists in which a symmetrical development is most probable (c). At very large distances, symmetrical development occurs again since the influence of the inhibitor is too small (e). The simulations are made under the assumption that, in addition to diffusion, some of the inhibitor is distributed instantaneously by convection.

distances lead preferentially to a reversal of orientation, and very large distances lead to symmetrical regenerates. In terms of the model, near-tip fragments can contain a fraction of the original activator maxima. This will direct the regenerating maxima according to the original orientation. Therefore the probability of having normally oriented regenerates reflects the activator concentration remaining in the fragment. It shows the expected steep decrease with distance from the original tip. At a larger distance, the only clue for the regeneration of the activator pattern would be the graded inhibitor distribution. In such a fragment, the lowest inhibitor concentration is at the side opposite to the original tip side. It will regenerate with a reversed orientation analogous to the dorso-ventral reorientation of the dorsal half of a sea urchin embryo (Fig. 4.4). The probability of regeneration with reversed orientation is a measure for the inhibitor gradient remaining in the fragment. As shown in Fig. 4.7, the probabilities, as observed through experiments, for normal and reverted regeneration are distributed similarly to the activator and inhibitor in the model proposed. Though the substances are not yet isolated, the experimental interference and the following regulation allow one to determine their approximate distributions.

Periodic structures

The possibility for the generation of periodic structures by coupled biochemical reactions has been shown in the pioneering paper of Turing (1952). In the proposed theory, periodic structures can be formed if the total area becomes larger than the range of the inhibitor; a small baseline activator

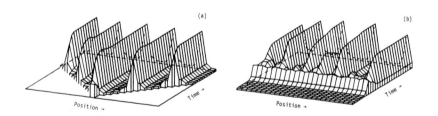

Fig. 4.8. Formation of periodic structures in space, Several activator maxima can emerge if the range of the inhibitor is smaller than the size of the field. A linear assembly of cells is assumed. (a) Regular spacing of peaks occur if pattern formation works during growth. New maxima appear in an area remote from existing maxima since the inhibition is too low there to suppress the onset of autocatalysis. Marginal growth at both ends is assumed and new maxima are added in the region of growth. (b) A somewhat irregular pattern arises if the pattern formation begins to work only after a certain size has been observed. The maxima first appear too close together and some are finally suppressed. A certain maximum and minimum distance is observed.

production (ρ_0 in eq. 3.2) can trigger a new activator maximum at a distance from an existing one. The resulting pattern will be fairly regular if the pattern formation mechanism has been working throughout the growth period, as shown in Fig. 4.8a. A new activator peak is formed whenever the distance to the nearest active centre exceeds a critical distance. Wilcox *et al.* (1973) have explained similarly the periodic appearance of heterocyst cells in the algae *Anabaena* by means of an induction-lateral inhibition mechanism. If the pattern formation is initiated by random fluctuation and has begun only after certain extension has been obtained, the spacing will be somewhat irregular (Figs 4.8b and 4.9); local maxima initially appear too close to one another, since the inhibition originating from the incipient centres is initially small. With increasing activator concentration, the mutual inhibition also increases and, therefore, some of the initially present activator peaks are, in the course of time, suppressed. An irregular spacing arises, but a maximum and minimum spacing is none the less observed. An example of a more irregular pattern is the spacing of cilia on the epidermis of *Xenopus* embryos and of the stomata, the apertures in the epidermis of leaves which are used for gas

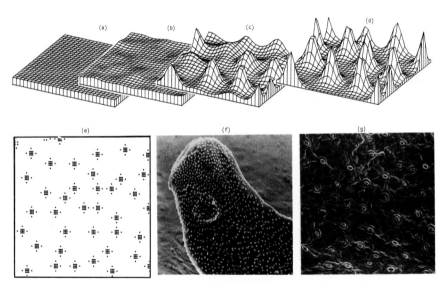

Fig. 4.9. Two-dimensional periodic pattern. (a–d) Stages in the formation of a bristle-like pattern in a non-growing field. The resulting pattern is irregular but a maximum and minimum distance is maintained. (e) A different calculation viewed from the top; the activator concentration is indicated by the density of dots. Biological examples: (f) the pattern of cilia on the surface of a *Xenopus* embryo (see Landström, 1977) and (g) of the stomata on a leaf (*Heleborus niger*, scanning electron micrographs kindly provided by M. Claviez).

exchange (Fig. 4.9f,g). Bünning and Sagromsky (1948) pointed out that the stomata formation begins only after the leaves have obtained a certain size and after the cell division in the epidermis has almost ceased, which implies that the concentration of a growth hormone has dropped below a critical level. However, cell division once again occurs adjacent to the stomata. The growth hormone seems to be distributed similarly to the inhibitor in the proposed theory, suggesting that they may be identical: initially the formation of the activator peak is suppressed by a constitutive inhibitor (growth hormone) production (ρ_1 in eq. 3.2) over the entire leaf. Only after the switching off of this production, activator peaks (signal for stomata formation) are formed with an irregular spacing. Each peak is surrounded by

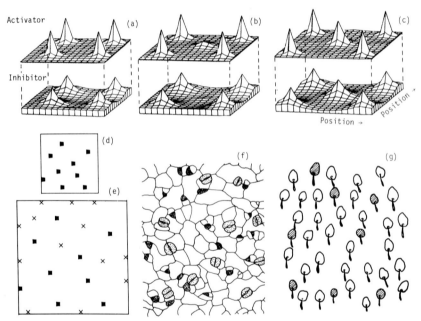

Fig. 4.10. Insertion of new activator maxima during intercalary growth. With increasing distances of the existing maxima, the inhibitor concentration can become low enough in some cells to allow the formation of new maxima. (a–c) Activator (top) and inhibitor distribution (bottom) in a two-dimensional, intercalary growing field. (d) Randomly initiated maxima (similar as in Fig. 4.9) and (e), newly inserted maxima (x) after intercalary growth. (f, g) Biological examples: (f) the epidermis of the leaf of *Alliaria* (Bünning and Sagromsky, 1948). The older stomata (pairs of dark cells) have obtained some distance from each other and new stomata (one dark cell) are initiated in-between. (g) Bristles and the surrounding plaques on the cuticle of the bug *Rhodnius* (Wigglesworth, 1940). During growth from the 4th to the 5th instar, new plaques (shaded) are formed in-between the older ones.

a cloud of inhibition (growth hormone)—leading to further cell divisions here. Later, the leaf grows further by expansion of the cells. If the stomata become too remote from one another, new stomata are formed at optimal spacing between the existing ones. Figures 4.10 and 4.11 show the insertion of new centres according to the theory.

The bristles and hairs on some bugs represent similarly a somewhat irregular pattern (Wigglesworth, 1940). To explain the actual distances, Lawrence (1966b, 1970) concluded that the inhibitory circles have no fixed extension but that some normally distributed range of extension is required. According to our theory, the spacing will naturally show some variations. Additional activator centres arise if the inhibitor concentration is below a certain level and not at a fixed distance from other centres. More remote centres will also contribute to the inhibitor concentration and have, therefore, an influence on the spacing. Since a fully developed activator peak

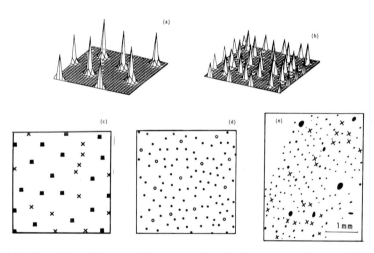

Fig. 4.11. Insertion of new activator maxima during progressive restriction of diffusion. (a, b) During a shrinkage of the diffusion range, the inhibition can become so low that new maxima are formed at locations distant to existing activator maxima (similar as during intercalary growth, Fig. 4.10). The maxima are lower since the inhibitor escapes much slower into the environment. (c) The new activator maxima (x) may appear quite close together in a former "activator valley". Biological examples. (d) Distribution of hairs (●) and sensory bristles (○) on the sternite of the bug *Oncopeltus* (after Lawrence, 1966b, 1970). During the first larval stages, sensory bristles (○) are formed which are remote from each other. During the fifth stage, hairs are formed which keep distance from each other and from the bristles. The reduced activator maxima (b, a) may be the signal to form hairs instead of bristles. (e) The distribution of chromatophores on the skin of a squid. The new chromatophores (x) can appear in groups between the older ones (●) (drawing kindly provided by A. Packard).

is surrounded by a high inhibitor concentration, a new centre can be formed only at a large distance from an existing one, whereas the initially lower inhibitor concentration of two simultaneously arising peaks allows their formation in closer proximity.

There are indications that in leaves, as well as in the insect epidermis, the spacing of different structures is controlled presumably by only one inhibitor; in leaves these structures are stomata and hairs, in insects (*Oncopeltus*)

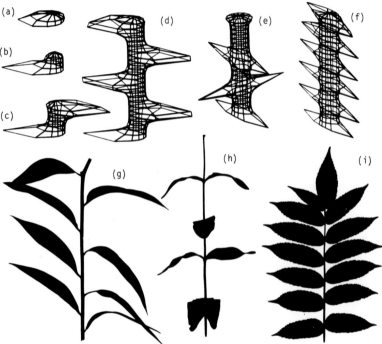

Fig. 4.12. Regular spacing of activator peaks as a model for phyllotaxis. (a–f) A growing shoot is simulated by doubling the cells at the upper end of a cylinder; random fluctuation may determine the location of the first maximum (a) which can be used as a signal to initiate a leaf. After further growth (b), (c) the next maximum appears on the opposite side due to the inhibition emanating from the first maximum; the final result (d) is an alternate (distichous) arrangement. Opposite (decussate) pattern (e) is formed if the diameter of the cylinder is higher or the diffusion range of the inhibitor is smaller, and especially if an inhibitory influence from the apex prevents new centres from arising near the apex. Parallel-arranged activator maxima (f) are formed if the growth is fast enough so that cells have some memory that their ancestors were originally activated or if the diffusion of the activator is facilitated in an axial direction. Examples of alternate and opposite leaf arrangements are given in (g) and (h). The parallelly arranged leaflets shown in (i) may arise from an activator pattern as shown in (f).

bristles and hairs (Fig. 4.11d). In *Oncopeltus*, bristles are formed only during the first four larval stages; hairs are determined in the last, or fifth, larval stage. Lawrence (1970) concluded that the distribution of hairs among the bristles is obtained from a shrinkage of the inhibitory fields. According to our theory, after such a shrinkage the newly formed peaks would be of considerably lower peak height (Fig. 4.11a,b) which may signify the change from the signal "make bristles" to "make hairs". During growth of a field or shrinkage of the inhibitor range, several activator maxima can appear quite close to each other in a former activator valley, arranged like pearls on a string (Fig. 4.11c). The younger chromomeres on a squid resemble such an arrangement (Fig. 4.11e).

A beautiful example of a very regular periodic pattern is the spacing of leaves—the phyllotaxis. The leaf primordia are formed during cell proliferation in the shoot apical meristem. There are currently two major approaches for the explanation of this phenomenon. One model supposes that leaf primordia are formed at the "first available space" (Iterson, 1907; Adler, 1974). Experiments involving surgical intervention (Richards, 1948; Snow and Snow, 1937; Wardlaw and Cutter, 1956) or treatment with plant hormones (Schwabe, 1971) support the second model which assumes a field of inhibition around each existing primordium and that new primordia are formed where the total inhibitory influence is least (Schoute, 1913). That is exactly the behaviour of the activator-inhibitor model. Figure 4.10 shows the insertion of a new maximum between existing maxima whose spacing increased due to intercalary growth. At the area of lowest inhibitor concentration, the inhibitor distribution is necessarily shallow and therefore several cells start with activator production. The emerging maximum sharpens itself, since inhibitor is produced by all these cells, a competition starts and only the best-located group will win. The resulting new maximum has the same size and shape as the others and is surrounded by its own inhibitory field. Figure 4.12 shows a simulation of a growing shoot by approximation to a growing cylinder. Depending on parameters, alternate, opposite, or parallel arrangements of activator peaks are formed which can initiate leaf formation. The helical arrangement of leaves and the Fibonacci series can be explained on the basis of lateral inhibition as well (Richter and Schranner, 1978; Mitchison, 1977). Newly formed buds in hydra (Fig. 6.1) also show alternate arrangement.

5

Polarity, size regulation and alternative molecular realization

Activator-depleted substrate model

The long-range inhibitory effect need not come from a physically existing substance but can be derived from a depletion of a substance necessary to the activator production. Depletion has been postulated as the inhibitory mechanism in tubularia (Morgan, 1904; Barth, 1940) and in the spacing of insect bristles (Wigglesworth, 1940). A possible interaction which leads, according to the theory (Gierer and Meinhardt, 1972), to pattern formation is given in eq. 5.1. It has similarities with the "Brusselator" proposed by Prigogine and Lefever (1968) but is simpler:

$$\frac{\partial a}{\partial t} = ca^2 s - \mu a + D_a \frac{\partial^2 a}{\partial x^2} \tag{5.1a}$$

$$\frac{\partial s}{\partial t} = c_0 - ca^2 s - vs + D_s \frac{\partial^2 s}{\partial x^2}. \tag{5.1b}$$

In this set of equations, the autocatalytic feature of activator production is maintained. The antagonistic effect results from the depletion of a highly diffusible substrate (s) which is consumed in the autocatalysis.

This mechanism has some properties different from the activator-inhibitor system which may allow an experimental distinction. In an activator-inhibitor system, an induction of a secondary activator peak is possible by an unspecific decrease of the inhibitor, e.g. by UV treatment or cell poisoning. In contrast, in an activator-depleted substrate interaction, neither the removal of the activator nor the removal of substrate will induce a new activation. The unspecific induction indicates the existence of a real inhibitor.

A characteristic feature of the activator-depleted substrate interaction is that the location of an established maximum can be shifted towards higher substrate concentrations. Figure 5.1 shows a simulation in a growing field. At an early stage of growth, high activator concentration is formed at one boundary. With increasing total size, the concentration of the substrate in the non-activated area increases. With that, the concentration of the substrate in the near vicinity of the activator maximum increases and also becomes steeper; the activator production near the maximum may therefore become higher compared to that at the maximum itself. Thus, the location of the activated area at a boundary becomes unstable and shifts into the centre of the area, where the substrate may be supplied from both sides of the activator peak. Therefore, such a mechanism is able to detect the centre of a field. With still further growth, the activator maximum would split into two, and both maxima separating from each other. Such a process may be at work in the determination of the central growth zone in a bacterial cell. An *E. coli*

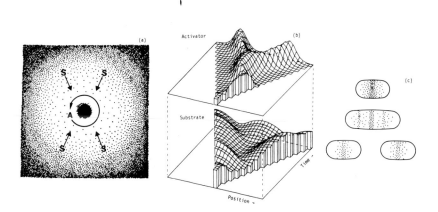

Fig. 5.1. Activator-depleted substrate model. (a) An even simpler pattern-forming reaction does not require an inhibitor. The autocatalysis of the activator (A) is antagonized by the depletion of a substance (s), for instance a precursor, which is used up in the autocatalysis (eq. 5.1). (b) Simulation in a linear array of cells, growing at both margins. When the field is small, only an activation at one margin is possible and gradients are formed. During growth, the substrate concentration in the non-activated part can increase so that the location of the activator peak becomes unstable; a shift to the centre follows. After further growth, the central peak will be split and the resulting two maxima separated. This can subsequently repeat itself. Equation 5.1 is used with the following constants: $c = 0.01$, $\mu = 0.01$, $D_a = 0.025$, $c_0 = 0.01$, $v = 0.0$, $D_s = 0.4$. (c) This centre-finding mechanism may be involved in the pattern formation of an *E. coli* bacterium. New material (black dots) is incorporated into the envelope of the bacterium only at a central zone of growth (Schwarz *et al.*, 1975). After some growth, the zone splits, the bacterium divides and the two zones are again in the centre of the two new bacteria.

bacterium is surrounded by a firm envelope, the murein sacculus, which determines the shape of the cell (Schwarz et al., 1975). Autoradiographic studies have revealed that new material is incorporated only in a small band in the centre of the bacterium (Fig. 5.1). If the size of the bacterium surpasses a certain length, the growth zone first splits into two. Only thereafter, division is initiated and the two growth zones are again located at the new centres of the two cells. The signal to form pole caps may be a certain substrate concentration. The pattern formation within the bacterium indicates that an autocatalytic step is involved in the growth of a bacterial envelope and that the supply of precursor molecules is rate-limiting in this process.

Thus, in the activator-depleted substrate system, a regular distribution of activator maxima may be obtained by splittings and shifts of previously existing activator centres as the field size increases. This is in contrast to the activator-inhibitor system where whole new activator centres arise at a distance from previous centres.

Autocatalysis may result from an inhibition of an inhibition

It is very possible that more than two substances are involved in the pattern-forming reaction. In these cases autocatalysis and lateral inhibition may be hidden in more complex reaction schemes. The following reaction consists of three components and does not contain any autocatalytic term:

$$\frac{\partial a}{\partial t} = \frac{1}{b^2} - a \tag{5.2a}$$

$$\frac{\partial b}{\partial t} = \frac{c}{a} - b \tag{5.2b}$$

$$\frac{\partial c}{\partial t} = a - c + D_c \frac{\partial^2 c}{\partial x^2}, \tag{5.2c}$$

(all constants have been set arbitrarily to unity and Michaelis–Menten constants in the nominator are omitted for simplicity). The effective autocatalysis results from the mutual inhibition of the substances a and b. Assume a and b at an equilibrium. A small increase of a would lead to a decrease of b and a lowered b concentration would lead in turn to a further increase of the a production. By itself, such a system of two substances inhibiting each other would be bistable and either a high a or a high b concentration would be attained (see p. 110). These two components together can play the part of the activator in a pattern-forming system. In addition, it has to be ensured that if in one part of a field, for instance, a is dominating, in the remaining part b must be dominating. This can again be achieved by a substance of a high diffusion range. For instance, as shown in eq. 5.2c, a may be converted into the diffusible substance c which undermines the repression of b by a, since it is in competition with the a molecules. The example demonstrates how careful one has to be with the attribute "activation" or "inhibition". Formally, the substance c has an activating effect on the b production. In the complete reaction system, however, the long-ranging c is antagonistic to the autocatalysis and leads to stabilization. It has, therefore, the function of the inhibitor. Formally, the concentration difference between the two (short-ranging) substances inhibiting each other has the function of the activator, causing the local destabilization (Gierer, 1981a).

The sources of the activator and inhibitor and their polarity-determining influence on the pattern

So far, we have assumed that the tissue is initially homogeneous, and every cell is able to synthesize the activator with the same efficiency. In fact, it has been one of our aims to show the possibility of pattern formation out of an otherwise unstructured tissue. However, a biological fact is that most tissues have an intrinsic asymmetry, a polarity. Axial differentials in respiration, in oxidation-reduction reactions, in the permeability, or in electric potentials have been detected in protozoa, eggs, embryos, hydroids, and some algae (see Child, 1929, 1941). Further, tissue fragments regenerate removed parts at a predictable position. For instance, in tubularia, an open apical end regenerates a hydrant and this is independent of the position of the cut. However, a piece of a column with two open ends does not regenerate a new hydrant on both ends, but only at the apical end. Morgan (1904) concluded from this fact that a competition exists between both sites and that a systematic advantage exists for more apical parts. That systematic advantage he has called "polarity". In terms of the model, certain prerequisites may be necessary for the synthesis of the activator and inhibitor, for instance, particular messengers or enzymes, ribosomes, energy-rich substances, such as ATP, or the presence of a certain cell type to which the synthesis is possibly restricted. We have called these necessary components "sources". In a formal sense a source is analogous to a water faucet. Then the effect of activator and inhibitor concentration is to open or close the faucet permitting more or less activator to be released. Thus, the activator and inhibitor concentrations decide to what extent the sources are active. The distribution of differentiated cells or cell constituents is a relatively stable tissue property, and a change requires much more time than the change of an activator distribution. The source distribution is the (relatively stable) polarity-determining tissue property, since it determines the *orientation* of the activator slope. To say it again, it is not assumed as in other gradient models that a local source creates the gradient but that minor asymmetries in the source density distribution orient the pattern. The pattern itself is generated by autocatalysis and lateral inhibition. The assumption of a graded source density is required on the basis of biological observations and not because it is logically necessary for pattern formation.

The source densities enter into the equations simply as factors in the auto- and crosscatalytic terms ($\rho(x)$, $\rho'(x)$ (eq. 3.2 and 3.3)). The simulation in Fig. 5.2 shows that even a very shallow asymmetry orients a pattern and that the resulting pattern is independent of details in the source distribution. Depending on the type of interaction, the final activator concentration may (eq. 3.3) or may not (eq. 3.2) depend on the absolute source density. Its influence will become important for the simulation of transplantation experiments with hydra (Figs 6.2 and 6.3).

A graded source distribution has another very important consequence: it stabilizes the formation of only one activator maximum in fields of different sizes. An activator-inhibitor system on its own would form one, two or several maxima, depending sensitively on the size of the field (Fig. 4.1). In contrast, in a system with a graded source density, that activator maximum which arises at the area of highest source density dominates. It suppresses other possible maxima very efficiently (since it produces, for instance, more inhibitor). This allows, for example, fragments of hydras to vary by at least a factor five in size although only one head is formed. In tissues with a graded source density, the regeneration of an activator maximum will occur always

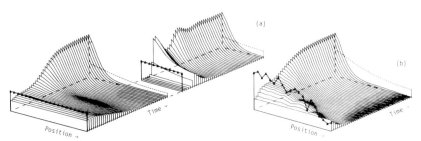

Fig. 5.2. Polarity determining influence of the source distribution on the emerging activator pattern. Any asymmetry in prerequisites necessary for the pattern-forming reaction—the sources—can orient the emerging activator pattern in a predictable way. In a reaction according to eq. 3.2, the final activator but not the inhibitor concentration (‒‒‒‒) is independent of the source density (▲‒‒▲‒‒▲‒). (a) Initiation of the pattern by a shallow source gradient. Removal of the activator maximum leads to its regeneration with the *same* polarity (compare with Fig. 4.4 and Fig. 4.7). (b) Higher absolute values or some fluctuations in the source density are without influence on the final activator pattern.

according to the original polarity. This is in contrast with regeneration in an isotropic tissue where an activator maximum can appear at the opposite end (Figs 4.4 and 4.7).

It may appear that a circular argument has been creeping in. We assume a graded source distribution to orient the activator gradient. But what is the origin of the source gradient? We have seen that in a growing area a monotonically graded activator distribution will appear even if the sources are homogenously distributed. If a high activator or inhibitor concentration causes a long term increase in the source density, the graded activator distribution leads to a more stable source distribution which provides the asymmetry to orient a regenerating pattern. The closed loop of a rapidly forming prepattern which generates a long-lasting asymmetry which can orient a prepattern, e.g. during regeneration, enables an infinite perpetuation. The asymmetry is maintained by a self-renewing process and is not diluted out if, for instance, a hydra is forced again and again to regenerate. In most biological cases, pattern formation does not involve symmetry breaking (although the proposed mechanism can perform this), since the tissue or its environment is asymmetric. The asymmetric organism forms an asymmetric egg and the orientation of the developing organism is therefore predictable.

Size regulation

The size of a particular substructure may be regulated in relation to the total size of the organism. For instance, during the formation of spores in the slime mould *Dictyostelium*, the front part of a migrating slug forms presumptive stalk cells, the remaining part forms prespore cells. The proportion of both cell types is well regulated over a large range of sizes of slugs (Fig. 5.3; Raper,

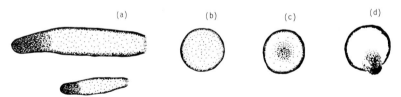

(a)   (b)   (c)   (d)

Fig. 5.3. Size-regulation in the slug of the slime mould *Dictyostelium discoideum*. The slug consists essentially of two cell types, the prestalk cells at the tip and the prespore cells at the rear. Vital stain is taken up in the vacuole of prestalk cells, they appear dark. (a) In slugs of different sizes, the ratio of prestalk prespore cells is maintained (see MacWilliams and Bonner, 1979). (b–d) Regeneration of a tip. After removal of the prestalk cells, new prestalk cells appear at random positions. A new tip area is formed by a chemotactic sorting out of the prestalk cells (b) 55, (c) 110 and (d) 160 min after tip removal (drawn after photographs kindly supplied by H. MacWilliams and A. Durston; see Durston and Vork, 1979).

1940; Bonner and Slifkin, 1949; MacWilliams and Bonner, 1979). Similarly, the size of the head of a hydra is regulated in relation of the total size of the animal (Bode and Bode, 1981).

The following mechanism would provide a means to measure the ratio of a particular cell type to the total number of cells. Let us assume two patches of differently determined cells, A and B. Only the A-cells produce a substance h which can diffuse freely between the A and B cells. The concentration of the substance h will be a measure for the ratio of A-cells to the total number of cells (A + B) since it is proportional to the A cells, the producers, and inversely proportional to the total number of cells (A + B) in which decay (and diffusion) takes place.

This method of size sensing can be combined with pattern formation if the activated cells (A-cells) produce the diffusible substance, the inhibitor h, in a fixed (more or less inhibitor-independent) amount. This can be achieved by a saturation of the activator production such as introduced in eq. 3.2 by the term $1/(1 + ka^2)$. With an increase of the total area, more space would be available into which the inhibitor can diffuse and decay. Due to the lowered inhibitor concentration, more and more cells switch from low to high but saturating activator production, balancing the ratio of activated to non-activated cells (Fig. 5.4). The saturation of the activator production has the consequence that the concentration maximum cannot increase in height but has to extend into a larger space. The activated area is therefore roughly proportional to the total area as long as the inhibitor range covers the total

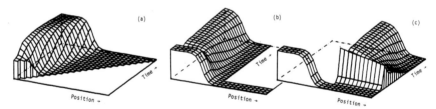

Fig. 5.4. (a–c) Size regulation of the activated region occurs if the maximum activator production is limited (eq. 3.2, $k \neq 0$). Due to the limitation, an increase of the total activator production can occur only by enlargement of the activator-producing area. The total activator production depends on the space into which the inhibitor can diffuse and decay. The activated region will be, therefore, roughly proportional to the total area as long as the inhibitor spreads out into the whole area by rapid diffusion or convection. (a) A linear array of cells, growing at the non-activated site. Due to size-regulation, the activated area increases in correct proportion. (b) After removal of the non-activated part from a distribution as shown in (a), the activated area shrinks until the corresponding size is obtained due to a build-up of inhibitor within the more confined space. (c) Removal of the activated part leads to a regeneration of the activated region corresponding to the smaller size.

field. An efficient inhibitor redistribution would be decisive for the range of size regulation. The size regulation also works when the pattern is already established, e.g. removal of the non-activated part leads to a shrinkage of the activated area (Fig. 5.4b). This size regulation would work as well in an activator-depleted substrate-scheme if the autocatalysis saturates. Then, all cells would produce a substance s which is consumed for the activation of cells and for the maintenance of this activation. Thus, the number of activated (s-consuming) cells will be in correct proportion to the total number of (s-producing) cells.

In contrast to the areas of high and of low activator concentration, the zone of transition between both areas does not adapt to the actual size (Fig. 5.4a) since it is determined by the diffusion range of the activator. The zone of transition is important for size regulation since a flip from high to low concentration or vice versa during expansion or shrinkage of the activated area occurs in this zone. A sharper zone of transition would increase the time required for adaptation to a new size. For very small areas, this transitional zone is too large to be neglected, and leads to an enlarged activated area. At the other extreme, field sizes larger than the range of inhibitor, the inhibitor cannot sense the total field and the activated areas formed are relatively too small. Both deviations from a perfect size regulation can be seen in *Dictyostelium* and in hydra (MacWilliams and Bonner, 1979; Bode and Bode, 1981).

The activated area can be maintained as a coherent area by a substantial activator diffusion. However, higher activator diffusion enlarges the zone of transition between high and low concentration and thereby deteriorates size regulation, as described above. A coherent activated area can be stabilized despite a small activator diffusion by a graded source distribution, since then the activation is strongly favoured in the area of high source density. In discussing the application of this model to *Dictyostelium*, MacWilliams and Bonner (1979) proposed that the experimentally observed sorting out of cells serves to set up such source gradient.

The slug of the slime mould: size regulation and pattern formation may be separated but coupled processes

In the model described so far, one and the same mechanism is able to generate a pattern and to control the size of the activated and non-activated portion. It may be satisfactory from the theoretical point of view to use only two substances for both purposes. However, very basic observations suggest that separate mechanisms are involved and good reasons can be given why this is of advantage.

The model would predict that in a size-regulated field a secondary maximum can hardly be induced since the inhibitor is distributed evenly in the whole field (otherwise a measurement of the total size would not be possible). Experimental observation contradicts that expectation. A large slime mould slug can decay into several smaller slugs and in hydra, near-head tissue can induce a second head (Fig. 6.2), indicating in both cases the ability to form secondary maxima.

Let us have a closer look at the slug of *Dictyostelium*. A distinction between future prestalk and prespore cells may occur very early after aggregation and possibly already during the aggregation phase itself (Maeda and Maeda, 1974). These cell types sort out in a chemotactic manner (Matsukuma and Durston, 1979; Durston and Vork, 1979; Tasaka and Takeuchi, 1981). This indicates that two different mechanisms are at work, one to form the correct number of prestalk and prespore cells and the other to position these cells correctly. In the model, if the activator is non-diffusable, the decision whether a particular cell will be activated is independent of the activation of the neighbour cells. The activated cells would be more or less randomly distributed among the non-activated cells (salt and pepper distribution), and the number of the activated cells will be in the correct proportion to the total number of cells. Which cells will become activated depends again on small differences in relative initial advantages described as source density. The fact that glucose-fed cells preferentially form prespore cells (Leach *et al.*, 1973) when aggregating together with non-glucose-fed cells can be explained in this way.

A second mechanism would be required to separate the intermingled prestalk and prespore cells. In *Dictyostelium*, this is achieved by a chemotactic movement of the prestalk cells towards the tip of the slug (Bonner, 1959; Durston and Vork, 1979). The signalling system for that chemotaxis may be similar as during the aggregation of amoeba and consist in a pulsative secretion of cAMP. However, any of the gradient forming mechanisms described above would be equally appropriate to generate an internal gradient. The high point of this gradient would form the tip of the slug. Both, prestalk and prespore cells, move uphill of this gradient and therefore the slug moves as a whole. To enable a sorting out of prestalk cells at the tip, the prestalk cells have to migrate faster than the prespore cells. It would be desirable that the centre of chemotactic attraction appears in a region of relatively high prestalk cell density. This occurs if the prestalk cells have a higher source density in respect to the chemotactic gradient system (see Fig. 5.2). With progressing chemotaxis and accumulation of prestalk cells at the tip, the tip will also be the region of highest source density. Thus, the prestalk–prespore system, on one hand, and the chemotactic gradient system, on the other, mutually reinforce each other. The gradient determines the tip which attracts the prestalk cells. The prestalk cells, in turn, are better

in producing the chemotactic signal and their collection at the tip stabilizes
the gradient.

The proportion-regulation in a regenerating slug (truncated *after* the
separation of the two cell populations) seems to be an argument against a
sorting-out mechanism. However, according to the model, proportion
regulation works also in regenerating slugs. For instance, after the removal of
prestalk cells (activated cells) at the tip, inhibitor is no longer produced in the
remaining slug. The inhibitor drops and new cells become autocatalytic.
Again, the cells with the relative highest source density become activated;
and they can be almost randomly distributed within the remaining field (Fig.
5.5). After completion of this first step (the selection of new prestalk cells),
the second process, the determination of the location of the new tip area,
follows. A new aggregation centre is formed at an area of the highest
density of the newly formed prestalk cells, attracting the other prestalk cells
(Fig. 5.3).

What is the advantage of having different mechanisms for pattern
formation and size regulation? If the two mechanisms were not separate,
pattern formation and, based on that, the preparatory phases of development into stalk and spore cells could start only after the slug has been formed.
Since the slug has a substantial size (1·5 mm), the life-time of the inhibitor
must be of the order of several hours to allow a complete equilibration of the
inhibitor and a correct measurement of the total size. Therefore, pattern
formation itself would require several hours and only then the signal would

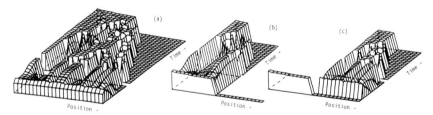

Fig. 5.5. Simulation of the size-regulation for *Dictyostelium* according to the two step
hypothesis. (a) If in a size-regulating activator-inhibitor system the activator
diffusion is very low and the source density shows only random fluctuations, the
activated cells are not located in a continuous patch of cells but are randomly
distributed. However, the ratio of activated/non-activated cells is independent of the
total number of cells. A separate mechanism (e.g. a chemotactic sorting-out of cells)
is assumed which separates the two populations. (b, c) This size-regulation also
works in regenerating slugs. If prespore cells are removed (b), the inhibitor accumulates temporarily since it can no longer escape into the non-activated part. This leads to
a conversion of some prestalk cells into prespore cells (deactivation). After removal
of the activated prestalk cells, the inhibitor drops (not shown) and new activated cells
arise (c), again initially randomly distributed (see Fig. 5.3).

be available for cells to develop into prestalk or prespore cells. Such a long time interval is especially disadvantageous for cells which have been running out of food, as is the case for the slug. The separation into two processes avoids these problems. Whenever cells aggregate, communication between neighbouring cells causes that about every third cell to become activated and to receive the signal "be prepared to become a stalk cell". That process can take place independently in many small aggregates and long before the final slug is formed. It is a fast process since it requires communication only between neighbouring cells. The gradient for the chemotactic sorting out can be formed simultaneously.

Some factors have been partially purified which are presumably involved in the pattern formation as described above. The differentiation-inducing factor DIF (Town *et al.*, 1976; Gross *et al.*, 1981) may be the substance *s* controlling the number of prestalk cells. The slug turning factor STF (Fisher *et al.*, 1981) may be the inhibitor in the gradient system which orients the chemotactic movement of the cells. Together with other factors influencing slug morphogenesis such as cAMP or ammonia (Sussman and Schindler, 1978), both substances may provide inroads for the biochemical characterization of a relatively simple pattern-forming process.

The argument for a separation of both processes in hydra would be similar. It is essential for a regenerating hydra to have a rapid decision over which group of cells should form the new head. The control of the final size is not such a critical decision and can consume more time.

Oscillating patterns and their use in chemotactic-sensitive cells

An activator peak, once formed, is very stable, and a shift in space by small external influences is nearly impossible, since an activator peak is, for instance, shielded by a "cloud" of inhibition. This stability is desirable for many situations; for others it is not. As already mentioned, small external influences can direct the location of a developing activator maximum. This mechanism offers a possibility for a chemotactic-sensitive cell to detect weak concentration differences. But the stability of an activator peak once formed would prevent a continuous adaptation to the changing external conditions. A possible solution would be an oscillating establishment and decay of an activator peak, whereby at each oscillation, the activator peak can be newly localized at the best position available. Periodic formation of the activator peak is obtained if the lifetime of the inhibitor is longer than that of the activator (Meinhardt and Gierer, 1974) or, in the activator-depleted substrate model, if the substrate concentration equilibrates too slowly. The periodic appearance of an activator maximum and the adaptation to a changed external gradient is shown in Fig. 5.6. The inhibitor, due to its longer lifetime, accumulates. From a certain level onwards the activator production is switched off and the remnant activator decays. After the decay of the inhibitor, a new activator maximum is triggered, starting either from an internal basic production or by an external supply. The location at which the maximum appears depends on small external influences and may be changed from one oscillation maximum to the next.

An example where an oscillating maximum could be used for the detection of an external gradient may be the aggregating amoeba of the slime mould *Dictyostelium discoideum* (see Gerisch, 1968; Loomis, 1975), which find each other by chemotaxis after a shortage of food. In the model, the cells would have a very sensitive phase for the orientation of the activator peak shortly before the next activation occurs. The short lifetime of the activator within the cell may reflect a secretion of the activator into the medium, thus providing a signal for other cells. The longer lifetime of the inhibitor (or the

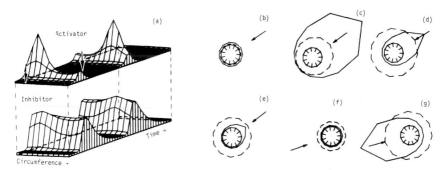

Fig. 5.6. A model for chemotactic sensitivity. The high amplification of small local difference which is inherent in the activator-lateral inhibition mechanism, can be used to detect the direction of an external gradient. A periodic adaptation to changing environmental conditions is possible if the activator concentration oscillates, as is demonstrated here with the activator-inhibitor model. Oscillation occurs if the inhibitor adapts too slowly to a changing activator concentration. The calculation is made for a circular object; to allow a space-time plot (a), the circle is cut up, the left and right ends of the represented distributions are, therefore, in reality adjacent neighbours. Some stages in the polar (intracellular) activation are shown in (b–g). The cell membrane is represented by a circular arrangement of elements, and the activator (——) and inhibitor distribution (– – –) is indicated. A shallow external gradient (2 % difference in ρ_0 across the cell, arrow points to the highest concentration on the cell surface) triggers a sharp local maximum which could then be used as a signal to draw out pseudopods at this side. After the (slow) decay of the inhibitor the next activation is possible. If the direction of the external gradient has been changed (arrow), the location of the activator peak will change accordingly. Such a mechanism may be the basis for the chemotactic sensitivity of aggregating amoeba of the slime mould *Dictyostelium discoideum*.

time necessary to restore the substrate concentration) causes a lag phase before the next oscillation can be stimulated. Increasing activator concentration in the medium can shorten the time until the next oscillation occurs which allows the synchronization of the individual cells. Therefore, in addition to the ability to respond in a directional way, this simple mechanism has all the properties demanded by Cohen and Robertson (1971) for the relay mechanism in *Dictyostelium*. The autocatalytic substance in the oscillation of *Dictyostelium* is cyclic AMP (Gerisch and Hess, 1974; Malchow *et al.*, 1978). The application of a small amount of cAMP to a suspension of cells can lead to a 100-fold increase of cAMP in comparison to the added amount. The pathway of the autocatalysis and of the antagonistic reaction which causes the oscillation is not yet clear.

A similar process may be going on in the growth cone of an extending nerve fibre. The growth cone has to find a particular larget cell, presumably by following a signalling gradient. Harrison (1910) has shown that nerve extension is not a continuous process but proceeds stepwise. The reason for this may be that nerve growth consists of two separate phases. First the cell senses the local gradient, leading to internal amplification of the signal. Following this period of amplification, the cell grows in the direction of the internal maximum, somewhat "blind" to the external gradient. More details about the formation of netlike structures will be given below (Chapter 15).

Strategies for the isolation of activators and inhibitors and expected pitfalls

For the maintenance of an inhomogeneous activator and inhibitor distribution, a continuous production and decay of both substances is necessary; this requires energy. In a developing egg the food supply is restricted. It is, therefore, to be expected that the cells produce the substances only in small quantities and that the cells are sufficiently sensitive. To give an example, the slime mould *Dictyostelium discoideum* is sensitive to a cAMP increase of as low as $3 \cdot 10^{-11}$ M (Malchow *et al.*, 1978). In cases in which higher concentrations of a substance are required for a change of development, it is likely that non-specific effects are observed.

The non-trivial antagonistic interaction between the autocatalytic and inhibitory substances is presumably the reason why no complete pattern-forming reaction is yet known. The general problem can be illustrated by an analogy. Let us assume that we would like to detect the mechanism whereby a few people become rich and others remain poor. Perhaps, for this investigation, we could arrange for a complete separation of all the rich people from the rest of the population. Instead of finding the mechanism, the so separated rich people will become poor and new rich ones will emerge. The property we sought to investigate has disappeared, since its maintenance requires a permanent interaction with the environment. Correspondingly, isolated small activated areas would lose their activation quite soon, since the inhibitor can no longer spread out and, in this way, the activation itself would be suppressed. In contrast, in the tissue thought to contain no activator, an intensive activator production starts almost immediately. The experimental task to separate activated and non-activated tissue fragments as initial step in the isolation of the substances would be incorrectly regarded as failure.

Biological tests are necessary to isolate both substances, and attention should be called to some precautions. The application of one of the substances can mimic the effect of the other. If, for instance, activator is added in a quantity which is large compared to that which occurs naturally, the activator is elevated everywhere. Due to the cross-catalysis, the inhibitor concentration will also increase tremendously. This suppresses the endogeneous activator production. In other words, externally supplied activator will not enhance but destroy the existing pattern. After the removal of the activator excess, the pattern is quickly restored, perhaps with a secondary activator peak at a new location. The same result would have been observed if inhibitor were supplied in large quantities. The local application of the activator or inhibitor at the active centre, as well as the removal of inhibitor there (see Fig. 8.4j), would suppress the formation of a secondary activation. A local application of activator at a distance from the activated centre can induce a second centre, but one has to take care that the effect does not arise from an unspecific loss of the inhibitor resulting from the treatment.

Perhaps the best strategy for the isolation of substances is to look first for inhibitory substances. For instance, if a substance suppresses regeneration at low concentration and if it is non-toxic for the tissue, the chance is high that the substance is involved in the normal morphogenetic process. Putative activating substances can then be tested to determine if they are able to overcome a low level of inhibition. Overall application of activating substances can be only successful if a quite strong inhomogeneity exists in the tissue; a local application would be better. If the existing activator centre has first been removed, the system would be especially sensitive to local addition of activator or inhibitor. If a periodic structure regenerates—as in the case of a hydra head with its tentacles—small amounts of activator may alter the number of the maxima.

No doubt, the isolation of both substances is a difficult task but we hope that the theoretical insights help in the design of assay systems, and especially in the interpretation of the results.

6

Almost a summary: hydra as a model organism

In higher organisms, the pattern-forming process leading to the primary subdivision of an organism occurs only once and in a short time interval. In many cases, it is then very difficult to gain access to the embryo for an experimental manipulation. In contrast, small pieces of the freshwater polyp hydra can regenerate the basic body pattern every time. It is therefore a convenient model organism to study the generation of a primary pattern and the diverse features of its pattern regulation allow a comparison of the proposed model and the real biological system (Gierer, 1977a).

A hydra is about 1–2 mm long (Fig. 6.1), contains about 100 000 cells but has only about seven different classes of cell types. It is essentially a cylinder consisting of an ectodermal and an entodermal cell layer. Characteristic structures are the hypostome with the opening of the gastric column for food uptake, the four to six tentacles, one or several buds and the basal disc to adhere to some support. The hydra has been used for the study of developmental processes for a long time. The ability to regenerate missing parts has been mentioned as early as 1744 by Abraham Trembley.

Despite the formation of new heads, many other features of hydra morphogenesis have been experimentally investigated, for instance the control of foot regeneration (MacWilliams et al., 1970), the determination of nerve cells from interstitial cells (Berking, 1979b), the regulation of multipotent stem cells (Bode and David, 1978), the change in the cell shape of evaginating buds (Graf and Gierer, 1980), the morphogenesis of nerve and stem cell free hydras (Campell, 1976). Mutants are available in which normal development is altered (Sugiyama, 1982), for instance, mutants, which cannot regenerate or which form multiple heads along the body column. The experimental investigation of the parameters which are changed in these mutants will provide more information about how normal development is controlled

48

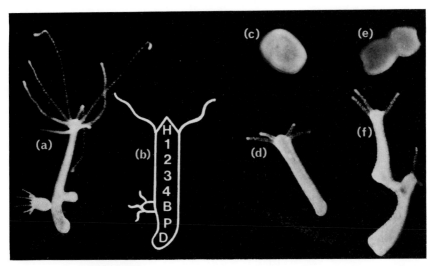

Fig. 6.1. Head regeneration in hydra. (a) The animal; (b) schematic drawing to illustrate the terminology H = head or hypostome, 1–4 gastric column, B = bud, P = peduncle and D = basal disc. (c–d) a regenerating 123 piece after 6 and 48h. (e, f) A 12123 graft (e) regenerates two heads (f) (see Fig. 6.2d).

(Sugiyama, 1982). With all this information (for a recent survey see Tardent and Tardent, 1980), the hydra has a good chance to provide us in the near future with a rather complete picture about how the development of a relatively simple organism is controlled.

All these features have to be incorporated to achieve a complete theoretical description of hydra, which has not yet been done. In the following section we will demonstrate that an application of the theory as developed so far in this book can account in great detail for very refined experiments concerning the formation of new heads.

The appearance of new heads after grafting operations

A very instructive set of axial grafts on hydra have been carried out by Wolpert *et al.* (1971). To have some reference points for the discussion of the graft experiments, they coined names for the parts of the hydra: the head (H), the gastric column (1, 2, 3, 4), the bud region (B), the peduncle (P) and the basal disc (D). In Fig. 6.2, the main results are sketched together with computer simulation on the basis of eq. 3.2, showing that the model is able to account for the experimental results with one set of parameters such as decay and diffusion rates. For instance, a combination of two hydra fragments H12/1234 leads very rarely to the formation of a new head at the site of

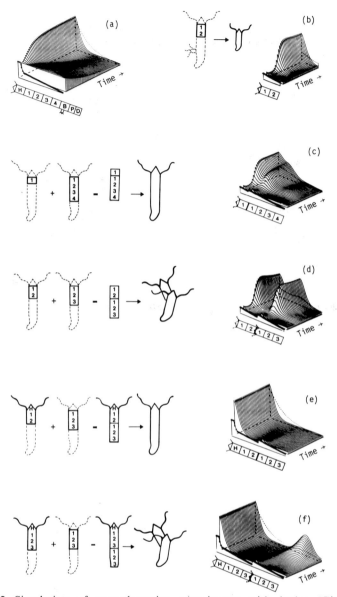

Fig. 6.2. Simulation of transplantation experiments with hydra (Gierer and Meinhardt, 1972, 1974). The simulation is intended to demonstrate the polarizing effect of a shallow source distribution and the suppression of an activator peak if induced too close to another one. High activator concentration is assumed to be the signal for head formation. The nerve cells are assumed to be the sources of activator and inhibitor (Schaller and Gierer, 1973; in eq. 3.2 $\rho = \rho'$); the assumed distribution is indicated (——●——●——). (a) Formation of the graded activator concentration which is used as the initial concentration before experimental interference in (b–f) and in Fig. 6.3. (b) Regeneration of a 1–2 piece. The source distribution assures regeneration in

connection while after removal of an existing head, e.g. in a combination 12/1234 a new head is frequently formed (Fig. 6.2e,d). Generally, tissue located originally close to the head has a strong tendency to form a head when implanted sufficiently remote from a head (Fig. 6.2f).

In terms of the model, a high activator concentration could be the signal to form the head. The local maximum is necessarily surrounded by a zone of inhibition. After head removal, the inhibitor drops and a new maximum will be formed in the remaining tissue, leading to the regeneration of the head. By the transplantation, a stimulus for triggering a new maximum is formed. The stimulus can consist of the remaining activator and/or of the step in the source density. The latter is presumably more important since the source density is more stable and lasts for a longer time. The probability of overcoming the local inhibition increases with increasing distance from the existing head (area of inhibitor production), but also when the transplanted tissue is derived from a location closer to the head, since then the resulting step in the source density is higher.

The polarity reversal experiments of Wilby and Webster (1970a,b, Fig. 6.3) provide a direct indication for the stability of the source density and for the time required for the inhibitor to diffuse through the body column. Removal of a head and grafting it at the end of the body column (1234H) leads to a head regeneration at the 1-piece. This proceeds like a normal regeneration since the inhibitor needs too much time to diffuse through the body column and the formation of a new prepattern cannot be suppressed. However, if the additional head is first grafted (H1234H) and later, after at least four hours, the original head is removed (1234H), sufficient time is available for the inhibitor to diffuse through the body column and the regeneration of the 1-piece can be suppressed (Fig. 6.3b). Such a "reversed" hydra is stable. However, a fragment derived from such an artificially reversed hydra regenerates the head according to the *original* polarity (Fig. 6.3c). This is a strong indication of a long-lasting tissue property with an orienting effect on the prepattern. It shows also that the activator pattern and the polarity-determining tissue property can be experimentally brought into conflict. As a rule, an established activator maximum will dominate over an oppositely

---

the original polarity, even if the activator distribution is completely uniform. (c) A small piece from near the head, 1, grafted onto a body column regenerates only one head due to the lateral inhibition. (d) A larger piece, 1–2, grafted onto the body column develops two heads. (e) The second head can be suppressed if the original head is left on the first piece. (f) If the first piece is longer and, therefore, the distance between the head and the site of the graft (containing a source density discontinuity) higher, the inhibition of the head may be too low and a new head can be formed. This simulation agrees with the experiments of Wolpert *et al.* (1971) and Hicklin *et al.* (1973).

oriented source distribution, while in the regeneration of an activator maximum, the source distribution will be decisive. If the reversed arrangement lasts for more than four days, the intrinsic polarity changes and regeneration will also occur at the end of the body column next to the transplanted head. In the long term, the prepattern imposes its orientation onto the tissue polarity.

The model is based upon the assumption that the polarity of the tissue results from a graded tissue composition and not from the alignment of polar

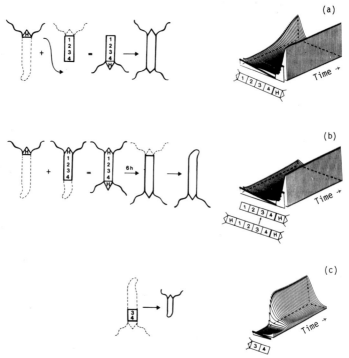

Fig. 6.3. Polarity reversal experiments in hydra (Wilby and Webster, 1970a,b) and their simulations. The experiments indicate the time requirement for the diffusion of the inhibitor through the animal and the stability of the source distribution. (a) If the head is removed and grafted at the opposite end of the body column, the inhibitor needs too much time to diffuse through the animal; regeneration of a head will take place. (b) If, in a similar experiment, the original head is removed no earlier than 4 hours after implantation of the second head, there is enough time available for the inhibitor to diffuse through the animal and, after removal of the original head, a regeneration is suppressed. (c) If the head of such a hydra with apparently reversed polarity is removed, the regeneration appears according to the original polarity; the source distribution, not the remnant activator concentration, is decisive for the location of the new activation.

cells. A dissociation-reaggregation experiment with hydra tissue (Gierer *et al.*, 1972) has provided direct evidence for this. In this experiment (Fig. 6.4) tissue derived from a near-head-region (H) and from the body column (C) has been dissociated. In this procedure, the orientation of the individual cells is certainly lost while the tissue composition remains unchanged. In composite reaggregates formed from clusters of H and C cells, most of the tentacles are formed from the H-cluster, indicating that the original position of the cells are decisive, as expected from the graded source density hypothesis. A computer simulation is shown in Fig. 6.4.

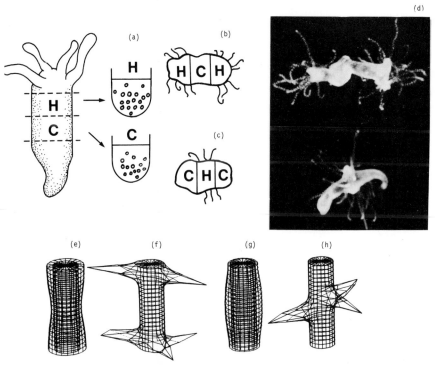

Fig. 6.4. Experimental evidence that polarity of hydra is based on cell composition and not on the alignment of individual polar cells (Gierer *et al.*, 1972). (a) Tissues derived from near the head (H) and from a more central region of the body column (C) has been dissociated. A possible alignment of the cells is certainly lost in this procedure. By centrifugation in a tube, sausage-like aggregates are produced, the H-cells are located either marginally (HCH) (b) or centrally (CHC) (c). After two days, new tentacles are formed mainly from the cells derived from near the head (H) (d). This preparation avoids any vital staining which could disturb the result. (e–h) Simulation in a cylindrical array of cells. (e, g) Assumed source density (difference between inner and outer cylinder); (f, h) final activator distribution. Maxima are formed at the margins or in the centre.

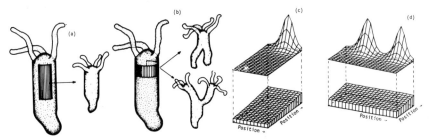

Fig. 6.5. Influence of the orientation of the source gradient on the number of heads formed. (a) Longitudinal strips of the body column regenerate only one head or foot. (b) In contrast, circumferential stripes regenerate, as a rule, two heads or two feet (Bode and Bode, 1980). (c, d) Simulation: a source gradient (bottom) oriented along the long side of a rectangular piece of tissue (c) leads to the formation of only one activator maximum (top) since the distance between the cells with high source density is small. (d) A strip of the same size and the same slope in the source density gradient but with a long extension perpendicular to the source density gradient forms two maxima. The total difference in the source density is smaller and cells with the same source density are much further apart. The "headstart" in the competition for head formation is thus less pronounced and two maxima can result.

Strips cut from a hydra cylinder regenerate differently depending on whether the stripes have an axial or circumferential orientation (Bode and Bode, 1980). In the latter case, two heads or feet are formed frequently (Fig. 6.5). The asymmetry-providing source density is graded in the axial but constant in the circumferential extension. In a circumferential strip the absolute differences in source densities are much smaller. Therefore cells of higher source density are further apart. The polarizing asymmetry of the tissue is less pronounced. During the competition between regions for making the head-signal, two maxima can arise.

Substances influencing hydra morphogenesis

Some factors are known which influence development of a hydra. Berking (1977, 1979a) has partially purified a substance which can inhibit bud formation as well as head and foot regeneration. The tissue is very sensitive to this inhibitor, concentrations of $< 10^{-9}$ M are sufficient to show detectable effects. The chemical composition of the inhibitor is not yet known, it is small (500–1000 DA) and it is not a peptide. Schaller (1973, 1981) found a "head activator" which increases the number of tentacles and enhances the transition from interstitial cells (stem cells) into nerve cells. It is a small peptide consisting of ten amino-acids, of which the sequence is known (Schaller and Bodenmüller, 1981). In terms of the model, this substance could be responsible for the feedback of the prepattern on the source density. It leads to an increase of the density of nerve cells which are, in turn, the main source of this head activator. Thus, the autocatalysis is indirect. From the point of the model, it would be reasonable to assume that the head activator with its determinative influence on nerve cell formation plays an essential role in the size-regulation of the head, but not in the determination of the position where the head is formed. The reasons why these two processes are separated have been given on p. 41. It could well be that a second head activator exists which determines the position of the head. A very helpful circumstance has facilitated the isolation of both these substances: they are stored in large amounts in a bound form. In hydra, it seems to be more the release of stored morphogenetic substances than the rate of synthesis which is decisive for the control of development. This appears reasonable since after head removal, the animal has to starve until a new head is formed. Head regeneration must be as fast as possible and can proceed only on the expense of available material.

7
Spatial sequences of structures under the control of a morphogen gradient

The concentration pattern formed by autocatalysis and lateral inhibition is assumed to be the signal or prepattern, to initiate a particular structure, for instance, the head of a hydra. Frequently, several structures are formed in a precise spatial relationship. They are determined under a common developmental control. Even in hydra it is not only the hypostome, the cone-shaped opening of the gastric column, which is formed during a regeneration but this structure is also surrounded by a ring of tentacles. More evident examples for sequences of structures are the digits of a vertebrate limb, the segments of an insect or an insect leg or the structures within such segments.

Intercalating versus non-intercalating sequences

Some of these sequences have the ability to regenerate missing elements by intercalary regeneration while others are unable to do so. Examples for both can be found in the development of insects. In cockroaches, an internal part removed from a particular leg segment will regenerate (Bohn, 1965; French, 1978). However, confrontation of different leg segments does not necessarily lead to the regeneration of the missing segments (Bohn, 1970a,b). Similarly, gaps in the basic body pattern of insects are not repaired. For instance, gaps induced by a temporal ligation of an *Euscelis* egg remain permanently (Armbruster and Sander, quoted after Sander, 1975b) and asymmetric bicaudal embryos of *Drosophila* (Fig. 8.3) develop such gaps without any experimental interference (Nüsslein-Volhard, 1977).

In looking for differences between systems which show and which do not show intercalary regeneration, it is remarkable that most non-intercalating sequences are determined in a period without much change in the geometry. The segments of insects are determined in the non-growing egg and little cell proliferation takes place during the critical period of blastoderm formation. Similarly, in regenerating insect legs, the newly formed sequence of segments is laid down in a very minute scale and proceeds without cell divisions. Only later, after the leg segments are already distinguishable, are these structures enlarged by growth (Bulliere, 1972; see Fig. 9.6). In the terminology of Morgan (1901), both these processes áre presumably morphallactic processes. Further, non-intercalating sequences have in many cases a clear organizing centre. In the chicken wing bud, a small group of cells, the so-called ZPA (see Fig. 10.7) is decisive for the formation of the digits. In some insects, a small area at the posterior pole of the egg, the activation centre, has to be present for a normal development (see Fig. 8.1). Frequently after an experimental interference a sequence of structures up to the most terminal structures are formed, e.g. UV irradiation can induce an additional insect abdomen (see Fig. 8.4), or an imperfect wound healing of an insect leg can lead to two new distal leg parts (see Fig. 9.7).

In contrast, systems which show intercalary regeneration seem to depend much less on special organizing centres but rather on an interaction between neighbouring cells (or groups of cells) at each level of the sequence, detecting and repairing any discontinuity. While in non-intercalating systems the tendency for a unidirectional proximo-distal or antero-posterior transformation exists, in intercalating systems there are mainly the distal element which regenerate the missing parts in a distal-proximal transformation (see Fig. 13.1; Bohn, 1972; French, 1976a; Nübler-Jung, 1977). On the basis of these differences one should expect that both processes are controlled by different mechanisms. Two possibilities can be envisaged for the generation of sequences. On the one hand, the local concentration of a gradedly distributed substance, the morphogen, determine the particular structure at the particular location. In terms of Wolpert (1969, 1971) it is the positional information and its interpretation which causes the sequence. The second possibility consists of a mutal induction of neighbouring structures. Explicit models of both types will be given and comparison with biological systems will reveal that positional information and its interpretation can account for the determination of segmented structures such as the insect body, the insect legs or digits of vertebrates. In contrast, the pattern formation within insect segments seems to be of the mutual induction type (p. 138). The differences listed above in the ability to intercalate and in the requirement for an organizing region will find straightforward explanations in these models.

Sources, sinks and the shape of the gradients

It has long been argued (Boveri, 1901; Child, 1929, 1941) that spatial organization could be accomplished by the graded distribution of substances termed morphogens. Wolpert (1969, 1971) developed this idea further into the concept of positional information. He pointed out that the size of an embryonic system is small when determination occurs, of the order of 1 mm or 100 cells across. Diffusion combined with local production and destruction at opposite ends, can form a gradient of this size within a few hours (Crick, 1970). This order of magnitude seems reasonable. Gradient formation involving diffusion in an area with a dimension of the order of 1 cm, on the other hand, would require a full day. It is thus tempting to speculate that the spatial development of an organism or of parts of it is controlled by a graded distribution of a substance during a stage of development where the extension of the region is of the order of 1 mm or less, and that the local concentration determines the further developmental pathway of each cell.

There are several ways to set up a graded distribution. The assumption of a morphogen source and/or sink alone would only shift the problem of morphogenesis to another level as long as no explanation is provided as to how and where they arise in an initially undifferentiated tissue. The mechanism of short-range autocatalysis and long-range inhibition described above can provide local high concentrations which may act as sources or sinks, and, in addition, explains why they usually appear at the boundaries of the system. A linear gradient can be obtained with a source at one side and a sink at the other (Crick, 1970), but such an arrangement has some disadvantages. First, the size regulation is poor; the concentration around the source depends on the distance between the source and the sink (Fig. 7.1a) unless there is a homeostatic mechanism by which the source strength is increased for smaller sizes. As a physical analogy, if a house has thinner walls, it requires more heating to maintain the same internal temperature, or the same inside-outside temperature gradient. Secondly, a linear gradient signifies that the relative morphogen increase per unit length is high at low morphogen concentration and low at high morphogen concentration. Therefore, the cell must be able to measure high concentrations much more precisely than low concentrations if it is to achieve the same spatial accuracy throughout the tissue. Both problems are avoided if the morphogen decays not only at the terminal sink but, to some extent, everywhere (Fig. 7.1b). The concentration around the source is then nearly independent of the total size and determined mainly by the local decay rate. The slope is steeper in the area of high morphogen concentration with an approximately constant change per unit length, as would be desired for uniform reading accuracy. In

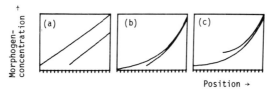

Fig. 7.1. (a–c) Concentration profiles of morphogenetic substances, generated by different source-sink arrangements and their dependence on field size. (a) A linear gradient can be formed by a source (here at the right side) and a sink (left). The concentration around the source will depend very much on the distance between the source and the sink, as demonstrated by the long and the short curve. (b) A uniform breakdown in addition to that at the terminal sink leads to an approximately size-independent concentration at the source and at the sink, since the concentration around the source depends more on the local decay rate. The relative concentration increase per unit length is approximately constant which facilitates the discrimination between different concentrations during the interpretation of positional information by the cells. (a) A simpler system consists only of a terminal source and a uniform breakdown. The advantage of simplicity is counterbalanced by the problem that, at the end opposite to the source, the concentration is size-dependent and shallow. Nevertheless, the central and source-containing portion of the gradient can be used to supply positional information; an asymmetric location of the determination of the structures is then to be expected. Positional information produced in this way seems to be used in the determination of segments in insects or of the digits in chicken limbs (Fig. 10.7).

addition, the time required to reach the steady-state concentration is much shorter. Therefore, much more convenient for cell specification than a linear gradient produced by a morphogen source and sink is an approximately exponential gradient formed by a terminal source with an overall decay of the morphogen. A local sink at the other boundary would maintain the morphogen at a low level, improving in this way the size-regulation.

In sea urchins, both the animal and the vegetal pole is of importance. Removal of vegetal cells leads to an "animalization" of the embryo and vice versa. As an explanation, Runnström (1929) has put forward the double gradient hypothesis. However, a source, overall degradation plus local sink system would explain the data as well. Let us assume a source at the vegetal and a sink at the animal pole. Removal of the source region, especially of the micromeres would lead to a general decrease of the morphogen and therefore yield an animalized embryo. The other way round, removal of the sink region (cells at the animal pole) would lead to a general increase of morphogen and therefore to embryos of the vegetal type (enlargement of the entoderm). A combination of the animal half (sink) and micromeres (source) forms a complete larva, despite the fact that most cells of the ventral half are missing since a source-sink combination can show a good size regulation (Fig. 7.1b).

Lithium ions have a vegetatizing effect. Parts of the animal portion of an embryo after culture for some time in a medium containing lithium ions can act in a similar way as the most vegetal cells, the micromeres (see Hörstadius, 1973). In terms of the model, lithium causes a general increase in the source strength. In contrast, rhodamine ions act as an animalizing agent, which is expected to poison the source.

The question arises if, in addition to the homogeneous destruction of the morphogen, a terminal sink is required at all. For the formation of a local sink a separate activator-inhibitor system would be necessary. A relatively simple pattern-forming system would, therefore, consist of a local source only and a uniformly distributed decay of the morphogen but without a local sink. The price paid for this simplicity is that the gradient at the end which does not contain the source is shallow (assuming the boundaries are impermeable), and the absolute concentration here is size-dependent (Fig. 7.1c). Appropriate positional information can be supplied only in the central region and that portion of the tissue containing the source of the gradient. In other words, only a certain concentration range of the gradient can be used. An advantage of using only a fraction of the gradient is that the mechanism then becomes insensitive to a size variation of the tissue over a certain range, since the fraction of the gradient used will be present both in a larger and a smaller field (Fig. 7.1c). Indeed, if only one organizing centre is involved, the area opposite to the organizing centre seems, in most cases, not to be used for the specification of structures. Two examples are the determination process in early insect development and the determination of the digits in chicken limb buds (Tickle et al., 1975), which will be discussed in detail below. The unused cells in the portion of tissue where the gradient is beyond the limit utilized for the relevant development may become necrotic and the constituent material recycled into the growing tissue. Or, vice versa, the utilization of the full region between the terminal boundaries of a diffusible gradient is a first indication that both ends contain organizing centres.

8

A gradient model for early insect development

Early insect development is a very instructive system for studying the determination of several structures within one process (for review, see Sander, 1976; Counce, 1973). After fertilization, the dividing nuclei in the egg spread out into the cytoplasm (cleavage stage) and migrate finally to the egg periphery, coming to rest in a well-defined layer (syncytial blastoderm stage). Only then are cell walls formed between the nuclei, leading to the cellular blastoderm. The embryo proper—the germ band—is formed out of a fraction of this blastoderm. The segments of the larva are linearly arranged and become individually distinguishable during germ band formation. The final pattern can be experimentally disturbed by centrifugation, ligation, thermocauterization, puncture, or UV-irradiation. Since the egg is well supplied with nutritional substances, a development into recognizable structures is possible even after severe experimental disturbances. A large amount of experimental data have been accumulated for many different species, providing a challenging testing ground for any model.

There are pronounced differences between species, nevertheless, as a working hypothesis we will assume that the basic developmental control is similar in all species. Keeping this in mind, the results can be generalized in the following way:

(1) The basic body pattern is controlled from the posterior egg pole.
(2) An instability exists at the anterior pole to form an abdomen instead of a head.
(3) Gaps can be formed in the sequence of segments which are not repaired by intercalation.
(4) The cells respond to the developmental signals in a stepwise manner.
(5) Segmentation and giving individual segments a particular "name" are separate but interdependent processes.

For the positional information it would not matter whether the gradient runs antero-posteriorly or vice versa. Is it possible to distinguish between both possibilities without having substances carrying positional information yet biochemically identified? Experimental interference at both egg poles can have unexpected consequences. For instance, puncture or irradiation of the anterior pole of a *Smittia* egg can lead to the formation of an abdomen instead of a head (see Fig. 8.4) whereas in *Euscelis* the shift of posterior pole material can lead to up to three abdominal structures (see Fig. 8.2). Insect development has been assumed therefore to be controlled by anterior and posterior determinants (Kalthoff, 1976). It has been shown, however, that many irradiation, plasma shift and ligation experiments are explicable under the assumption of a gradient arising from the posterior pole alone (Meinhardt, 1977). The sensitivity of the anterior pole reflects more an instability against the formation of a second source. Since it is believed that the control of insect development is a paradigm for the control of development in general, this model should be described in some detail and compared with the experimental observations.

The "activation centre"—an organizer region at the posterior pole

Seidel (1929) found evidence in *Platycnemis* for an "activation centre", a small area at the posterior pole which is necessary for the organized development of the embryo. The fate-map indicates that this activation centre does not participate in the formation of the embryo proper; instead its duty is to organize the embryo. An exclusion of the posterior eighth of the egg by a ligation suppresses embryonic development (Fig. 8.1). However, the exclusion of only a slightly smaller posterior fragment leads to a normal development. If the operation is made early in the development, the result is either a completely normal development or no development at all; no intermediate forms are observed. No such centre can be detected at the anterior pole since a similar constriction there leads always to normal development.

If one assumes that the antero-posterior organization of the egg is accomplished by a morphogen gradient which is generated by an activator-inhibitor system, these experiments tell much about the orientation and shape of the distributions. Evidently, the autocatalytic centre must be localized at the posterior pole. The fact that no pattern regenerates after an early elimination of the posterior pole indicates that a small basic inhibitor production (activator-independent, ρ_1 in eq. 3.2) can suppress the auto-catalysis at very low activator concentrations. The smallness of the activation centre indicates that the activator maximum is very narrow, otherwise a regeneration of the pattern and therefore normal development would be expected even after removal of a much larger fragment. On the other hand, if the activator maximum is very sharp, the activator concentration is very low in almost the whole egg space and is therefore inconvenient for supply of positional information. However, any substance with a more shallow distribution, produced by the very localized activator maximum could act as

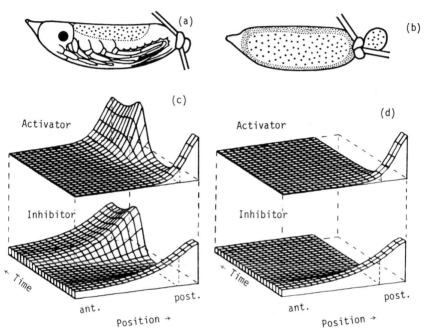

Fig. 8.1. The importance of the posterior pole in insect development. (a, b) Normal development of an embryo of a dragonfly is possible only if less than c. 1/8 of the posterior egg is excluded by an early ligation (Seidel, 1929) otherwise the blastoderm cells do not differentiate (b). A similar ligation at the anterior pole is without effect. (c, d) Simulation: in this and the following simulation, it is assumed that the positional information is generated by an activator (top)-inhibitor (bottom) system and that the distributions have attained a steady state (as shown in Fig. 4.1) during oogenesis. To show the reaction of the system to the experimental interference, both concentrations are plotted as functions of the antero-posterior position and time. (c) After removal of the activator maximum, regeneration can take place if sufficient activator remains in the egg to initiate the autocatalysis, restoring the gradient. (d) After complete removal of the activator maximum, its reformation depends on the small constitutive activator and inhibitor production (ρ_0 and ρ_1 in eq. 3.2). To maintain a monotonically graded distribution, secondary maxima have to be suppressed and this requires a low ρ_0 and/or high ρ_1. This can suppress the reformation of the removed maxima and no positional information would be supplied.

morphogen. Since the inhibitor production is activator-controlled and since the inhibitor has, due to its higher diffusion rate, a graded distribution throughout the total area, the inhibitor is a reasonable candidate for the morphogen. Our assumption will be, therefore, that a high activator concentration is formed at the posterior pole and that the cells or nuclei and their immediate plasma environment "learn" from the local inhibitor concentration which segment they must form. In this scheme, the inhibitor plays a dual role: it activates particular control genes and suppresses the formation of other activated areas.

The all-or-nothing effect after removal of the posterior fractions of the egg is easily explained on the basis of the model: either sufficient activator remains to overcome the basic inhibitor level and to reform the distributions via autocatalysis or all concentrations drop to a very low level (Fig. 8.1d).

In further experiments, Seidel (1935) removed large parts of the activation centre by burning with a hot needle. He was surprised by the result that even more than half of the posterior pole can be burnt and still yield normal development. With a gradual elimination of a constitutive source one would expect a gradual decrease in the morphogen concentration. But an autocatalytically activated source will restore the pattern as long as sufficient activator is available to initiate the autocatalysis.

Evidence for autocatalysis and lateral inhibition—pattern formation in leaf-hopper embryo *Euscelis*

More support for the positional information concept and for the organization from the posterior pole can be deduced from experiments with the eggs of the leaf-hopper *Euscelis* (Sander, 1959, 1960, 1961a,b). In this insect, a ball of symbionts is located at the posterior pole of the egg. These symbionts, bacteria necessary for the normal development of the embryo, are implanted in the egg by the mother. A dislocation of this posterior pole material in an anterior direction has a dramatic effect on the further development. After shift and ligation, up to three abdominal structures can be formed within one egg, some with reversed polarity. The main results are sketched in Fig. 8.2 and can be summarized as follows:

(1) After a shift of the symbionts and a ligation of the egg, partial embryos are formed in the posterior part. They have either a symmetric or a reversed arrangement of the segments.

(2) The anterior half of the blastoderm does not participate in the formation of the embryo proper. However, after an anterior shift of the symbionts and a delayed ligation, a complete embryo can be formed in the anterior fragment even if the symbionts are not included in that fragment.

Sander (1960) has concluded that the important factors are not the symbionts themselves but some "posterior pole material" spreading out from them.

Similarly, as in Seidel's experiment, one has to conclude that a small area controls the whole pattern formation in the antero-posterior dimension. The symbionts provide a handle to manipulate this area. The experiments are explicable by the theory assuming that some activated plasma is shifted with

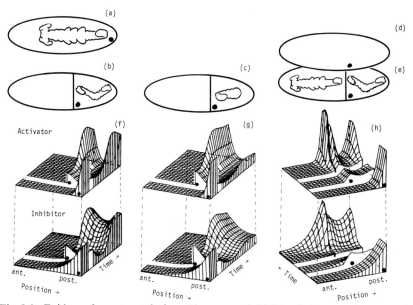

Fig. 8.2. Evidence for autocatalysis and long range inhibition in insect embryogenesis. (a–e) Experiments of Sander (1961a, 1962) with eggs of the leaf hopper *Euscelis*. Normal (a) and altered germ band patterns after shift of the posterior pole material (●) and ligation: either a symmetric (b) or a reverted sequence of abdominal segments (c) results. (d, e) If some time elapses between the shift (d) and ligation (e), a complete embryo is formed in the anterior fragment. Up to three abdomina can be formed within one egg. (f–h) Model calculation: shift of the symbionts is assumed to cause some redistribution of the activator. A break in the distributions indicates the time of an experimental interference. Despite the fact that the redistribution can be only vaguely controlled, only two new distributions are possible. Either (f) both maxima coexist keeping maximum distance from each other and a symmetric pattern emerges, corresponding to the result sketched in (b), or (g) the new maximum dominates over the old one via the long-ranging inhibitor, the resulting pattern has a reversed polarity. (h) If some time elapses between the redistribution of activator and the ligation, the anterior part is "infected" with sufficient activator that, due to the autocatalysis of the activator, complete gradients are formed. This corresponds to the result (e). To have a convenient perspective, the distributions in (h) are rotated 90° (after Meinhardt. 1977).

the symbionts (Meinhardt, 1977). Traces of this activated plasma can develop a fully activated source, preferentially at the physical boundaries of the (ligated) egg. In contrast, a stable morphogen source subdivided into two or three parts would lead a much reduced maximum morphogen concentration and no abdominal structure would be expected—in contradiction to the experiment. The newly formed maximum can either be dominant over the old one, leading to a reversed morphogen distribution or both maxima can coexist, leading to the symmetric pattern (Fig. 8.2). Obviously, there is some ambiguity between polar and symmetric patterns after an experimental interference. This is also a property of the theory and therefore an explanation is given of why minor and uncontrollable differences can lead to the two strikingly different, but well defined, alternative patterns. If sufficient time elapses between shift and ligation, the newly formed maximum can spread out, the anterior portion becomes "infected" and a complete pattern can be formed there (Fig. 8.2h). This observation also supports the autocatalytic aspect of the theory.

Vogel (1978) has separated three fragments of *Euscelis* eggs by two ligations. He observed that the central fragment has to have a relatively large size if a single pattern element is to occur while some little additional space is sufficient to add further segments. This fits nicely into the model where a minimum extension (range of the activator) is required to form a pattern. Around this critical size, the concentration range depends sensitively upon the size of the field. The maximum concentration and the concentration range of the gradient may be reduced (see Fig. 4.1a). The concentration would not be high enough to form the complete abdomen. Then, the most posterior structures are thoracic structures, in agreement with Vogel's observation.

Formation of posterior structures at the anterior pole

After certain experimental interferences, many species form abdominal structures instead of head structures in their anterior portion. Frequently, a completely symmetric development is observed. The experimental treatments evoking such "double abdomen" (DA) malformations are quite diverse: UV-irradiation (Yajima, 1964; Kalthoff and Sander, 1968) or puncturing (Schmidt *et al.*, 1975) of the posterior pole, temporary ligation (van der Meer, 1978) and centrifugation (Yajima, 1960). Double abdomen formation has also been found in a maternal effect mutant of *Drosophila* (Bull, 1966; Nüsslein-Volhard, 1977).

In the model, the abdominal structures are formed where the inhibitor concentration is high. The formation of additional posterior structures at unusual locations would indicate the triggering of a second activator maximum, establishing a second morphogen source. An especially favourable location for the formation of a second activation is the anterior pole, since here the inhibitor has its lowest concentration and any unspecific reduction of the inhibitor concentration may be sufficient to induce a new centre of activation. This is in agreement with the unspecific modes of double abdomen (DA) induction already mentioned. The induction of a DA has similarities with the unspecific induction of a second amphibian embryo (Waddington et al., 1936). However, in amphibians, it is not the antero-posterior axis but the dorso-ventral axis which becomes duplicated, forming a dorso-ventral-dorsal pattern which leads to two parallel aligned embryos. After DA-formation in insects, the two embryos are not separated. Both gradients overlap because the two sources are not sufficiently remote from each other.

Figure 8.3 shows photographs of a normal and of DA (bicaudal) larvae of Drosophila. Important for the gradient model is that the anterior half of the embryo is not merely transformed into the posterior half. In the centre of a normal blastoderm, the metathorax is laid down (Lohs-Schardin et al., 1979) while in DA-embryos, the plane of symmetry is very variable but always located in one of the abdominal segments. Therefore, the fate map of

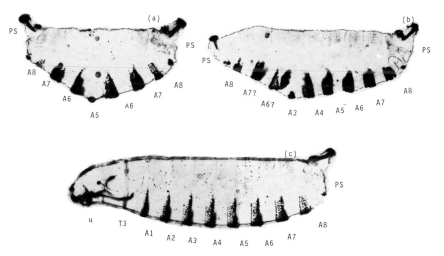

Fig. 8.3. The mutation bicaudal in Drosophila (Nüsslein-Volhard, 1977, 1979). (a) Mirror-symmetrical first instar larva with two abdomens. Head and thoracic structures are missing. (b) An asymmetric double-abdomen embryo. Since the terminal structures are present but the number of segments is different on both sides, one or two abdominal segments must be missing on one side. This gap is obviously not repaired by intercalation. (c) For comparison, a normal larva, A1 ... A8: abdominal segments; PS: the most posterior structure, the posterior spiracles; T3: metathorax; H: head. (Photographs kindly supplied by Nüsslein-Volhard.)

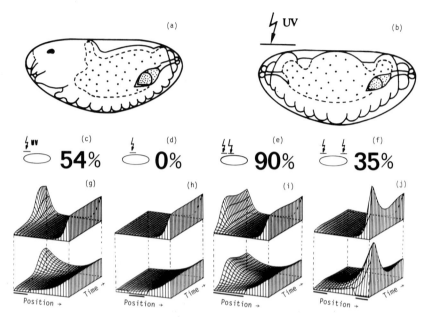

Fig. 8.4. Double abdomen (DA) formation in midge *Smittia*. (a) Normal embryo. (b) Irradiation of the anterior quarter of a *Smittia* egg can lead to a completely symmetric embryo with one abdomen at each pole (Kalthoff and Sander, 1968; drawn after Kalthoff, 1976). (c–f) Results of the experiments of Kalthoff (1971). The dose of the irradiation of the anterior quarter was adjusted to yield about 50% DA; the dose of additional irradiations was somewhat smaller, in order to minimize the number of eggs which fail to develop at all. The frequency of DA-formation is given in percent. (g–j) Model: the inhibitor is assumed to be UV-sensitive. Shown is the response of the activator (top)—inhibitor (bottom) system as function of position and time after irradiation. A reduction of the inhibitor concentration (black bar) at the anterior pole (g) allows an increase of the activator concentration which can, via autocatalysis, develop into a full second maximum. The inhibitor distribution (positional information) at the anterior pole becomes a mirror image of that of the posterior pole. Experiment: while an irradiation of the second anterior quarter is without effect (d), applied together with an irradiation of the first quarter, it considerably increases the probability of DA induction (e). Model: the removed inhibitor in a central area is rapidly replenished by the nearby source (h) and, therefore, without effect. But such a removal delays the restoration of the inhibitor concentration after an irradiation of the anterior quarter. Therefore, the activator increase after an irradiation of the anterior half is much more rapid (i) and the probability of reaching the critical level for the DA-formation is increased. Experiment: (f) an irradiation at the posterior pole is without serious effect (0% DA), but such an irradiation cures partly the anterior radiation damage. (j) Model: inhibitor reduction at the activated site leads to an overshoot of the activator and, consequently, also in the inhibitor concentration. This is without serious effect, since all concentrations necessary for the determination of any particular structure remain present. But the overall increased inhibitor concentration reduces the activator increase after irradiation of the anterior end; the activator increase may not be sufficient to reach the threshold for further autocatalysis and may, therefore, disappear.

far more than the half of the blastoderm is changed. As discussed below in more detail, this is expected from the overlap of two gradients and provides crucial support for the assumption of a diffusible signal.

In the midge *Smittia*, DA-formation can be induced by UV-irradiation (Kalthoff and Sander, 1968) or by a puncture (Schmidt *et al.*, 1975) at the anterior pole. According to the model, the UV-irradiation may either destroy the inhibitor or the inhibitor-producing structures. The results of a very instructive set of experiments by Kalthoff (1971) are shown in Fig. 8.4, together with their explanation in terms of the theory assuming that the inhibitor is UV-sensitive. Due to the inhibitor reduction, the activator concentration increases. If this activator increase is sufficiently high, a new activator maximum develops via autocatalysis, even if the inhibitor concentration is rapidly restored. If the activator concentration fails to reach the critical level, the activator increase will disappear. In agreement with the experiment, the formation of a second activation is an all-or-nothing event. Substantial support for the postulated interaction between an activator and a long-ranging inhibitor can be derived from the fact that a simultaneous posterior irradiation *reduces* the probability of induction of DA, so to speak, it cures the anterior radiation damage. A reduction of the inhibitor at the activated (posterior) site produces an overshoot of activator. As a result, more inhibitor is subsequently produced which spreads out quickly by diffusion and then acts to reduce the probability of triggering a second activation centre at the anterior pole (Fig. 8.4f,j).

After centrifugation of *Smittia* eggs, Rau and Kalthoff (1980) found embryos with double abdomen and double head formation. Most interestingly, they also find a complete reversal of the embryo in relation to the egg axis. In the latter case, the head is formed at the posterior egg pole while the abdomen is formed at the anterior egg pole. This observation fits nicely with the self-regulatory properties of the proposed mechanism and the possible shapes which the gradient can attain (see Fig. 4.1).

The long range character of the positional signal

The altered central segment in DA embryos offers a crucial support for the gradient model. From a central ligation of a *Smittia* egg at the blastoderm stage—at a stage when the segment pattern is fixed—we know that the segment No. 5, a thorax segment, is laid down in the centre (see Fig. 8.7). However, in double-abdomen embryos, the plane of symmetry at the centre is, as a rule, formed in segment 8 or 9. According to the model, in DA embryos, the morphogen diffuses from both sides into the centre and the concentration is thus increased there. This has the consequence that a more posterior structure is formed (Fig. 8.5). From the simulation of the early ligation experiments one can estimate the diffusion constants and lifetimes of the activator and inhibitor. Applying this to a simulation of a double abdomen formation leads to segment 9 as the central element, in essential agreement with the experiment. After an early ligation, the segment 9 is also formed in the posterior part of the egg (Fig. 8.7). The fact that the same element is formed in the centre after the two very different manipulations— anterior UV-irradiation and central ligation—is a straightforward consequence of the gradient model. The ligation renders impossible any flow through the centre and the morphogen accumulates. Similarly, after DA induction the flow at the centre is, due to the symmetry, also zero. Therefore, the same concentration and thus the same structure is expected after both manipulations. This is exactly what has been observed. This may be the best evidence available that the pattern is controlled by a long-ranging diffusible substance and not, for instance, by a chain of induction.

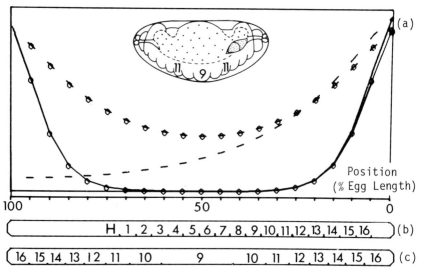

Fig. 8.5. Evidence for a long ranging morphogen. During normal development, segment 5, a thoracic segment is formed at the centre (see Fig. 8.7). In DA, the central element is around segment 8 or 9, an abdominal segment. The central segment is changed even though only a small anterior position has been irradiated. According to the model, after induction of a second activator maxima, inhibitor (– – –), diffuses from both sites towards the centre, causing an elevation of the inhibitor concentration there (positional information). This leads—far away from the site of experimental interference—to the determination of more posterior structures, in agreement with the experimental observation. (a) activator (——) and inhibitor (– – –) distribution in the normal and in DA embryo (○○○). (b) Fatemap of the normal embryo as derived from late ligation experiments (Fig. 8.7). (c) Calculated fatemap of a DA-embryo (after Meinhardt, 1977).

Negative size regulation—a phenomenon characteristic for gradient systems generated by a local source

In many developmental systems, the complete set of structures is formed even if a substantial portion of the developmental field is removed. Examples are the dorso-ventral axis of amphibians (p. 150) or insects (p. 130). However, the pattern regulation of the antero-posterior axis of insects exhibits the reverse behaviour. In a fragment of an egg, resulting from an early ligation (at the nuclear cleavage stage), *less* segments are formed then in the same area in an undisturbed egg, so to say, a negative size regulation. We will show that this is a straightforward consequence whenever a pattern is controlled by a morphogen gradient which is generated by a local source and diffusion.

Let us regard first only the posterior, or source-containing fragment of a ligated egg. In terms of the model, a ligation during the cleavage stage introduces a diffusion barrier. This leads to an accumulation of the morphogen. Due to the ligation and morphogen increase, a particular cell will get a more posterior specification: thus a particular structure will appear in a more anterior position. The fate maps of ligated and non-ligated eggs of *Drosophila* (Newman and Schubiger, 1980) have provided direct evidence for the predicted anterior shift. The segments which would have been determined just anteriorly of the ligation are instead omitted.

This feature of the model has counterparts in many experimental observations. Ligating an egg during the cleavage stage leads to an omission of segments, for example in *Euscelis* (Sander, 1959), *Calliphora* (Nitschmann, 1959), *Bruchidius* (Jung, 1966), *Protophormia* (Herth and Sander, 1973) or *Drosophila* (Schubiger and Wood, 1977). In *Euscelis*, for instance, a ligation at 44% EL (EL = egg length, 0% = posterior pole) (Fig. 8.6) at the blastoderm stage leads to a complete embryo in the posterior portion. The same ligation made earlier leads to an embryo lacking the head lobe. Figure 8.6 shows that this is expected from the accumulation of the morphogen. An early ligation placed more anteriorly, at 57% EL, leads to a complete embryo. In the enlarged space sufficiently low morphogen concentrations are possible despite the accumulation. The latter observation indicates that the incomplete embryo formed after the early 44% EL ligation does not result from an elimination of "anterior determinants", spreading out from the anterior egg pole since these influences would be eliminated by a 57% EL ligation as well. These experiments also rule out the argument that the omission of structures result from a damage of cells. As the late 44% ligation shows, the blastoderm cells located more anteriorly are normally not used in the embryo formation. The omission of segments does not result from a damage of cells but from a change in the positional information to which the cells are exposed.

The phenomenon of negative size-regulation also provides a strong argument against another type of model stipulating that the most posterior structure is induced at the posterior pole and that the more anterior segments result from a chain of induction, one triggering the next like falling pieces of a domino game. This sequential triggering must be assumed to be due to a more or less local process. Therefore pinching off regions which would not be used in the normal embryo genesis is expected to be without effect, in contrast to the experimental observation.

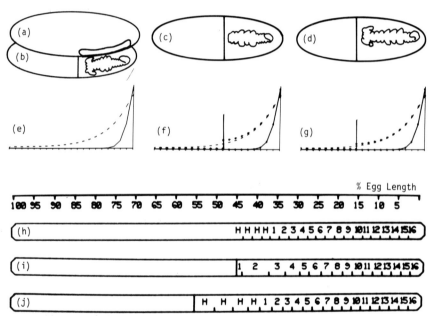

Fig. 8.6. Indication for a "negative" size regulation and the absence of anterior determinants in the leaf hopper *Euscelis* (Sander, 1959). (a, b) Only the posterior part of the blastoderm egg is used for the formation of the germband: a ligation at the blastoderm stage at 45% EL leads to the formation of a complete embryo (b). (c) An early ligation at the same position leads to an omission of the head. In the smaller area, less strucures are formed compared to the same region in the unoperated egg (negative size regulation). (d) A complete embryo is formed, however, in a larger fragment (57% EL), indicating that the omission of head structures does not result from the absence of anterior determinants. Model: (e) normal activator (——) and inhibitor (– – – –) distribution. (f, g) Due to the ligation, the inhibitor (+ + +) accumulates. The low concentration required for head formation is present only if the fragment is larger. (h–j) Assumed fate maps of normal (h) and calculated fate map of the ligated eggs (i, j).

The interpretation of positional information is a stepwise unidirectional and irreversible process

After a ligation of a *Smittia* egg (Fig. 8.7, Sander, 1975a) the terminal structures remain present but the more central segments, normally formed from blastoderm cells located on both sides of the ligation, are omitted. Note that the gap in the sequence of segments is asymmetric. If a ligation is performed more anteriorly, the segments omitted belong more to the

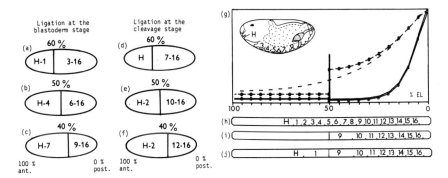

Fig. 8.7. The influence of a diffusion barrier and evidence for a stepwise, unidirectional interpretation of positional information. (a–f) Schematic drawing of the ligation experiments with eggs of the insect *Smittia* (Sander, 1975b). Inset in (g): Schematic drawing of a *Smittia* embryo. Segments are designated H, 1, 2, ... 16. (a–f) Experimentally observed germ band fragments after the ligation. The first and last segment formed in each germ band fragment is indicated. (a–c) After a ligation during the blastoderm stage very few—if any—segments are omitted: the egg behaves as a mosaic. This allows the drawing-up of an approximate fatemap (h). (d–f) If, however, the ligation is made earlier, during the cleavage stage, many segments are omitted, but the terminal segments always remain present. Model: after a ligation, the inhibitor (positional information, – – – –) accumulates on the source-containing posterior side (g), which leads to a shift of the segments in an anterior direction (i). Some of the segments normally determined posterior to the ligation would no longer be formed; in this example the elements 5–8 would be omitted (h, i). In the anterior portion, on the other hand, the inhibitor concentration decreases to a level which normally never occurs (g) and no element would be expected to be formed, which is at variance with the experimental observation (e). This contradiction has forced the assumption that the determination proceeds under the influence of the morphogen stepwise and unidirectional to more posterior structures until the determination corresponds to the local morphogen concentration. The structure formed in the anterior part after a ligation reveal how far the determination has already advanced at the time of the ligation (j). The pattern in the posterior portion depends on the extent of morphogen accumulation and the time available to adapt the determination to this increased morphogen concentration.

posterior portion, while in the anterior part, a head lobe is formed nearly independent of the time of operation. But if the ligation is performed at a more posterior location, more segments are lost in the posterior part than gained in the anterior part. From the experimentally known omission of segments in the posterior portion and the minimum time in which a second activation can be formed, one can estimate the diffusion rate of the inhibitor to be 5×10^{-9} cm^2 s^{-1} and its lifetime to be 1 h, respectively.

A description of the missing segments in the anterior portion with these parameters is not in agreement with the experiments. From the estimated short lifetime, one would expect a relative fast decay of the inhibitor (morphogen) such that no segments at all would be formed in the anterior part (Fig. 8.7). On the contrary, the head lobe is always formed. As Figs 8.7a and 8.7d show, a ligation of 60% EL of a *Smittia* egg leads to nearly the same segments being formed in the anterior part independent of whether the ligation is made early (during the cleavage) or late (during the blastoderm stage). In such a ligation experiment, a particular cell or nucleus seems to be determined at the time of the ligation if located anterior to the ligation, but the final pathway can be changed if located posterior to the ligation. Extensive reprogramming seems to be possible so that a more anterior structure can be reprogrammed to form a more posterior structure but not vice versa.

The explanation I have proposed for this stipulates how the cells measure the local morphogen concentration. Originally, all cells are programmed to form the most anterior structure. Under the influence of the morphogen, the cells proceed stepwise to higher (more posterior) determinations "head"— "thorax"—etc., until the determination corresponds to the local morphogen concentration. A step in the determination is—as in other developmental systems—essentially irreversible. If the morphogen disappears before the final determination is achieved, stepping through the different determinations will be interrupted. The determination would remain unchanged at the stage already reached. Such a situation exists in the anterior part after a ligation. If the morphogen increases due to experimental interference, for instance by the accumulation of the morphogen in the posterior portion after a ligation or after the induction of a second activation by UV irradiation, the determination can proceed. Structures corresponding to more posterior positions would then be formed. The omission of segments on both sites of a ligation results therefore from different reasons. On the posterior site, it depends on the accumulation of the morphogen; at the anterior site, it reveals how far the interpretation had progressed. This explains the asymmetry of the gap. At an early stage, when only the most anterior segments are already determined, a headlobe is formed in the anterior fragment, fairly independent of the position of the ligation (Fig. 8.7d,f). In the posterior fragment, the lowest morphogen concentration and therewith the most anterior structure depends essentially on the position of the ligation.

The type of stepwise and irreversible determination described above appears to be a general process. The proximo-distal determination of insect legs (p. 88) as well as the antero-posterior determination of vertebrate limbs (Fig. 10.7) follows the same rules.

Determination—or commitment—of a group of cells to form, say, a head lobe must consist of switching "on" a particular set of genes. Detailed models for the selection of gene activity under morphogen control will be given below (Fig. 11.5 and Chapter 14).

Alternative models

Other mechanisms which have been proposed for the control of insect development may appear reasonable as well. However, a look into their consequences reveals features which are not supported by the experiments mentioned above. Some of them are listed below, together with conflicting observations. Maybe, some of these conflicts may be cured by additional assumptions. The discussion should show how stringent the experimental observations really are for any model.

Model A. The prepattern specifies only the terminal element, the abdomen; the missing elements are specified by a chain of induction (such as shown in Fig. 13.3). Problem: a zone is expected in which the final determination takes place and which moves in a wave-like manner over the field. A ligation at a particular location should lead to different results depending on whether the wave has passed this position or not. If passed already, the development would be normal in both fragments and indistinguishable from a mosaic development. In contrast, when the ligation is made before the wave has passed this location, the development would be normal in the posterior fragment, but no development would be expected in the anterior fragment. In contrast, experimental observations show that gaps in the sequences of segments become gradually smaller if the ligation is made at a later developmental stage and that, especially in the posterior fragment, segments are missing also (Sander, 1976).

Model B. Two gradients, for instance *a* and *p*, with opposite orientations are formed by local sources on each end of the egg (anterior or posterior determinants). The ratio of both concentrations (a/p or p/a) is used as positional information (Sander, 1961a). Problems: after ligation, the concentration of each component drops to very low values in the fragments not containing its source. The values are lower than would be found anywhere in the normal embryo. Therefore the ratio would attain very high values on one side of the ligation and very low values on the other values which are out of the range used to specify structures in the undisturbed organism and this is clearly at variance with the observed gap behaviour.

Model C. Head and abdominal structures are determined by anterior or posterior determinants and missing structures are filled in by some sort of intercalation from both sides at the discontinuity. Problems: in a bicaudal embryo (see Fig. 8.3) or in a UV-induced double abdomen it has to be assumed that both ends bear posterior determinants. No discontinuity would be present to initiate the intercalation and therefore no pattern formation at all would be expected.

Model D. Similar to model C, but in the centre where neither anterior nor posterior determinants are present, thoracic structures are determined (Vogel, 1978) and again missing structures are made by intercalation. Problems: ligation through the centre should contain at each site thoracic structures and normal development is expected. Double abdomina should show thoracic structures in the centre. Neither of these expectations is in agreement with the experimental observations.

Model E. Specification of the field is achieved by a sequence of binary subdivisions (Kauffman *et al.*, 1978). A gradient-like prepattern bisects a field initially into two parts, anterior and posterior (0 and 1). By shrinkage of the diffusion range the prepattern changes into a bell-shaped distribution, subdividing the two halves into four quarters (00, 01, 11, 10), and so on. Problems:

(1) If part of the egg were removed, the pattern will be restored after corresponding shrinkage of the chemical wavelength. In a ligation experiment, such a mechanism would not lead to a gap but would produce a normal set of body parts in both halves of the egg. The only defect could be the absence of some fine structure if the shrinkage of the diffusion range remains insufficient. In contrast, the experiments show that each half produces even fewer structures than would be expected from mosaic development.

(2) At the blastoderm stage, the thoracic segments become subdivided into anterior and posterior compartments (Garcia-Bellido *et at.*, 1973, 1976). At that time wings and legs are not yet separated. On the basis of chemical wavelength one would expect an organization of the large dorso-ventral dimension first and only then a finer subdivision of the narrow segments into the even narrower anterior and posterior compartments.

(3) A cell which has seen a zero-concentration twice must be in a different state (state zero–zero) compared with a cell which has been exposed to "nothing, but only once" (state zero). This would require, for example, an additional counting or clock mechanism synchronized with the pattern-forming process. The relatively long-time interval in which a double abdomen can be induced in an insect (Ripley and Kalthoff, 1981) argues against a clock mechanism and against early irreversible binary decisions.

Open questions

The gradient model, despite providing a unified explanation for the many experimental observations, is not free of problems. Some observations which are difficult to integrate should be mentioned.

(1) After temporal ligation of eggs of the beetle *Callosobruchus*, van der Meer (1978) found a double abdomen formation in the right or the left half only while the other half was normal. Possibly, the cells in the non-affected half are unable to respond to the altered morphogen distribution.

(2) Centrifugation of eggs of *Smittia* (Rau and Kalthoff, 1980) and of *Chironomus* (Yajima, 1960) can lead to the formation of symmetrical double heads (or double cephalon) embryos which lack thoracic and abdominal structures. A similar pattern has been observed in a mutant of *Drosophila* (Lohs-Schardin and Sander, 1976). A central activator maximum is expected as one possible pattern after an experimental interference (see Fig. 4.1) but in such a case, two abdominal structures pointing with the posterior ends towards each other are expected. Such patterns do occur in an *Euscelis* egg after a shift of the posterior pole material (Fig. 8.2e) but only if the egg is also ligated. It seems that the dorsal and the ventral side of the egg have to be brought into contact to enable a high point for the antero-posterior organization. This would guarantee that the dorso-ventral and the antero-posterior pattern are oriented perpendicular to each other. An analogous observation has been made in planarians (see p. 136). Therefore, it may be difficult to induce a full height peak in the centre of an egg without a ligation. The symmetrical head-like structure mentioned above indicates a much reduced morphogen concentration.

(3) *Drosophila* embryos, despite using a large fraction of the blastoderm, show a remarkable insensitivity of the segment pattern against a variation of genetically altered egg size (Nüsslein-Volhard, 1979). The size-regulation seems to be connected with the formation of anterior structures because double abdomen embryos form indeed less segments if the egg is smaller. This size-regulation could be achieved by a somewhat stronger sink property of the anterior egg pole, keeping the morphogen concentration low (see Fig. 7.1).

9

Pattern formation in subfields: formation of new organizing regions by cooperation of compartments

In the preceding section, evidence has been presented that the positional information in an (insect) embryo is generated by autocatalysis and lateral inhibition. Under the influence of such a morphogen gradient, a subdivision into defined groups of differently determined cells is possible. The final spatial structure is, of course, much more complex than what would be achievable by the interpretation of one (or two orthogonal) gradients. Further subdivisions are clearly necessary. A possibility consists in the formation of secondary gradients and their subsequent interpretation. For example, a primary gradient may specify the future limb area and a secondary gradient can then specify the finer details of the limb, such as the digits. Detailed experimental data about pattern formation in developmental subfields are available for the imaginal discs of holometabolous insects and of the limb field in vertebrates. It will be shown that many of these experiments are explicable under the assumption that the boundaries between patches of differently determined cells, determined under the influence of the primary gradient(s), become the organizing regions for the developmental control of subfields (Meinhardt, 1980). Since the boundaries of existing structures give rise to the new structures, the existing and the new structures have necessarily the correct spatial relationship to each other. This allows a very reliable finer subdivision of a developing embryo.

Imaginal discs, their fate maps and compartment borders

Epithelial structures such as eyes, antennas, wings, halteres or legs are generated from nests of cells, the so-called imaginal discs (see Gehring and Nöthiger, 1973). In *Drosophila*, the cells of the imaginal discs are almost completely determined before pupation begins, at the end of the third larval stage. Fragments transplanted directly into metamorphosing larvae differentiate according to their original position within their disc. This allows a fate map of the disc to be constructed. Figure 9.1 shows a wing disc and some of the corresponding adult structures. In the leg disc, the leg primordia are arranged in concentric rings (Schubiger, 1968, see Fig. 9.2). The outer rings form the more proximal structures such as thorax and coxa, while the inner rings form the more distal structures such as tarsus and claws. The leg attains its final shape by a telescope-like extension of the central (distal) part.

Two features of the spatial determination of imaginal discs appear to be a key element in the understanding of how subpatterns are formed: their

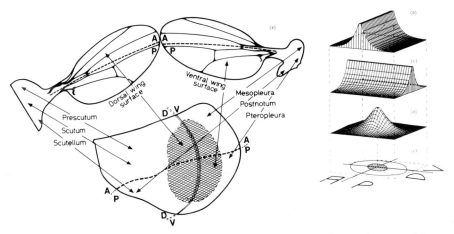

Fig. 9.1. The wing and its coordinate system. (a) The dorsal and ventral aspect of the wing. Below: the imaginal disc from which the corresponding adult structures are derived (after Bryant, 1975a; Garcia-Bellido *et al.*, 1976). The border between the anterior (A) and posterior (P) compartments does not coincide with any morphologically recognizable structure while the dorsal (D) and ventral (V) compartments form the corresponding wing surfaces as well as thoracic structures. (b–d) Model for the generation of the coordinate system. By cooperation of the A and the P compartment as well as of the D and the V compartment, two ridge-like morphogen profiles are generated (b, c). The symmetrical distributions are centred over the corresponding boundaries. The product of the A-P and D-V pattern has a cone-shaped distribution (d) which is appropriate to organize the proximo-distal axis. Only those cells exposed at least to a low threshold concentration become imaginal disc cells. Cells exposed to a relative high concentration form the wing blade (e). The primary event is therefore the formation of the boundaries. The imaginal disc is formed, in a secondary event, from cells surrounding the intersection of the AP and DV boundary. In agreement with the model, the distalmost structure, the wing tip, is formed at the intersection of compartment borders.

progressive compartmentalization (Garcia-Bellido *et al.*, 1973, 1976; Steiner, 1976; Crick and Lawrence, 1975) and the properties of pattern regulation (Schubiger, 1971; French *et al.*, 1976; Bryant, 1978). In this chapter we will develop a model about how compartmentalization and the generation of positional information in a subpattern are linked with one another. How compartments themselves can be formed is discussed elsewhere (p. 131 and Chapter 14).

One of the earliest developmental decisions is the separation into anterior and posterior compartments (Garcia-Bellido *et al.*, 1973, 1976: Steiner, 1976). It occurs during or shortly after blastoderm formation. A group of cells and all their progeny, once determined to form, for instance, the anterior part of the leg or wing will under normal circumstances never be reprogrammed to form a structure in the posterior part. They are thus said to be clonally restricted and a cell will never cross a precisely defined compartmental border. The separation of a thoracic segment into an

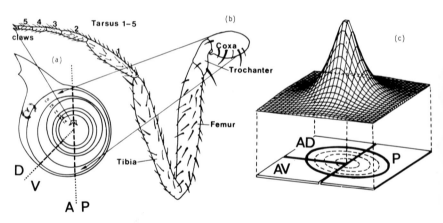

Fig. 9.2. The leg disc of *Drosophila*. Fate map, compartment borders and the proposed mechanism for the generation of positional information in the proximo-distal dimension. (a) The leg disc is subdivided into only three compartments (Steiner, 1976): the two anterior (anterior dorsal, AD, and anterior ventral, AV) and the posterior (P) compartment. The proximo-distal sequence of adult leg structure (b) map on the disc as concentric rings, the most distal structure is formed at the intersection of compartments. (c) The proposed explanation: the three compartments (AD, AV and P) cooperate to produce the morphogen. A high production rate is possible only where cells of all compartments are close to each other, at the intersection. The local concentration of the resulting cone-shaped morphogen distribution provides the positional information which dictates that the segment is formed. The concentric arrangement of the primordia is a straightforward consequence of this type of pattern formation. A certain minimum concentration is required so that a cell becomes a disc cell.

anterior and posterior compartment occurs almost simultaneously with the clonal separation of the segments themselves. The organization of the dorso-ventral axis of the thoracic segments follows a few hours later. In other words, the decision whether a cell obtains an anterior or posterior specification is made prior to the decision whether a cell participates in leg or wing formation (Wieschaus and Gehring, 1976; Steiner, 1976). This indicates that primarily the antero-posterior (A-P) border is formed and that, secondly, cells from both sides of the border are allocated to form the discs. It is not that existing discs become subdivided into the A and P compartment. Further subdivisions of the disc cells into yet other compartments follow.

Regeneration, duplication and distal transformation

Fragments of imaginal discs cultivated in the abdominal cavity of adult flies are capable of substantial pattern regulation. Two major types have been observed: either the structures remaining in the fragment are duplicated, leading to a mirror-symmetrical pattern; or the missing structures are regenerated. As a rule, when a disc is fragmented into two portions, the two fragments show complementary behaviour: one fragment duplicates itself while the other regenerates (Schubiger, 1971; van der Meer and Ouweneel, 1974; Bryant, 1975a,b). It has further proved possible to predict on the basis of its size and geometric position within a disc whether a given fragment will duplicate or regenerate. The borderline between the fragments which regenerate and those which duplicate does not coincide with any compart-ment border. For instance, a fragment containing constituents of only the anterior leg compartment can regenerate all the missing structures of the leg. This pattern regulation has been successfully described by the formal polar coordinate model of French et al. (1976). In this model it is stipulated that two positional parameters are of relevance for pattern regulation: position along the proximo-distal axis and circumferential position. Circumferential positional values (particular determined states) are assigned to structures around the circumference of a disc, arranged like the numbers of a clockface. Two rules are sufficient to describe how the pattern is regulated: (1) the shortest intercalation rule: whenever a fragment of a disc is removed, wound closure causes those cells at the wound surface to find themselves close to unusual neighbours. Missing structures are regenerated by intercalary regeneration according to the shortest intercalation rule. Only those structures which are necessary to reform a continuum will be regenerated. Whenever more than half of the positional values are removed, shortest intercalation leads to duplication; otherwise regeneration will occur. (2) The complete circle rule: distal transformation and outgrowth occurs whenever all the circumferential positional values are present, forming a complete circle. Duplicated structures (in which more than half the positional values are missing) are therefore not expected to show distal transformation.

As mentioned above, the borderline between fragments which regenerate and fragments which duplicate does not coincide with any compartment border. Compartments are therefore not an element of the polar coordinate model. Compartments are, however, the primary developmental subdivision of a disc. It will be shown how the two concepts can be linked.

Pattern formation by cooperation of compartments

In principle, a subpattern can be generated by autocatalysis and lateral inhibition as described in the preceding chapters. However, after interpretation of a primary gradient, sharp boundaries exist between the patches of differently determined cells (compartments). These boundaries open a new possibility for the generation of positional information. Let us assume two patches with a common boundary which cooperate for the production of a substance which acts as a morphogen. Due to the required cooperation, the synthesis of the morphogen is possibly only at the boundary. A symmetrical, ridge-like morphogen distribution centred over the boundary will result. Three compartments—similar as three countries—meet each other only at one point. If three compartments (or two pairs of compartments) have to cooperate, morphogen production is possible only at the point where cells of all compartmental specifications are close to each other. The point of intersection of the compartment borders becomes the source region of the morphogen. By diffusion and decay, a cone-shaped morphogen distribution is formed with the highest concentration at the intersection of the compartment borders (Fig. 9.2). The local concentration is a measure of the distance from the intersection and can be used as positional information in the proximo-distal dimension. The most distal structures are formed at the intersection of the compartments and the interpretation of the cone-shaped morphogen distribution leads in a straightforward manner to the circular arrangement of structures. The same morphogen distribution can determine which cells form the disc and which form the larval ectoderm. Only those cells exposed to a concentration above a certain threshold would participate in disc formation; no separate positional information system is required. According to this view, an imaginal disc never exists without subdivisions. The formation of the borders necessarily precedes the formation of the disc. Since the boundaries are determined under the influence of the primary organizing gradients, the emerging discs naturally have the correct orientation in respect to the body axis. The handedness of each disc is also determined since three patches, touching each other at one point, are sufficient to determine handedness in an unequivocal way.

Molecular mechanisms for such a cooperation are easily constructed. For instance, each compartment may be responsible for a particular step in the synthesis of the morphogen or each compartment may produce a diffusible co-factor which is required for morphogen production, Figs 9.2 and 9.4 have been calculated in this way. The positional information may be generated in two steps. By the cooperation of the A-P and of the D-V compartments two ridge-like distributions are formed which can supply positional information for the antero-posterior and the dorso-ventral dimension. The symmetrical distributions can be interpreted differently in the corresponding compartments, leading for instance to the partially symmetrical pattern of the wing. The product of the two ridge-like distributions then assumes the cone-shaped distribution (Fig. 9.1), organizing the proximo-distal dimension.

As discussed below, several lines of experimental evidence indicate that interpretation of proximo-distal positional information in discs proceeds in a stepwise, unidirectional manner, i.e. in the same way as in the early insect embryogenesis (Fig. 8.7). A distal determination, once obtained under the influence of the local morphogen concentration, seems to be irreversible. The "complete circle rule" for distal transformation of French *et al.* (1976) which is difficult to interpret in molecular terms, is thus simplified to yield the straightforward mechanism of "cooperation of compartments". The achievements of the complete circle rule, such as explanation of supernumerary appendages, remain valid, since demanding a complete circle is formally equivalent to requiring that cells of three or four sectors are close to each other.

The model links early compartmentalization and generation of positional information in the proximo-distal dimension. Two stipulations are made: cooperation of compartments in formation of a cone-shaped morphogen distribution, and response of the cells in a stepwise, unidirectional manner. Both assumptions are supported by experimental observations. It should be pointed out that the local morphogen concentration determines only, for instance, which leg segment a group of cells has to form. The fine structure *within* a segment is assumed to be generated by a different process (Chapter 13).

Evidence for the cooperation of compartments in the generation of positional information

The most distal structures are formed at the intersections of the major compartment borders. The tip of the wing is determined at the location where the A-P and the D-V border cross each other (Fig. 9.1). In the leg disc, the precise location of the D-V border at the centre is not known. However, the most distal structures, the two claws, are located on both sides of the A-P border and the D-V border points in that direction (Fig. 9.2). The most distal structures are not located trivially at the centre of the disc, since the posterior compartment is smaller. This is also true in earlier stages; the posterior compartment is made up of about half as many founder cells as is the anterior compartment (Garcia-Bellido *et al.*, 1973).

Distal transformation of leg fragments requires a close juxtaposition of all compartmental specifications. As can be seen from the experiments of Schubiger and Schubiger (1978) and Strub (1977a) the upper lateral quarter of a leg disc fragment (Fig. 9.3f) does not regenerate the removed distal primordia (centre of the disc). It does not contain the ventral compartment. Similarly, the lower medial quarter (Fig. 9.3b) contains the anterior-dorsal compartment only marginally and shows a low frequency of distal transformation. In contrast, a fragment which contains cells of all compartmental specifications shows distal transformation very frequently (Fig. 9.3d).

A complete set of circumferential structures is not required for distal transformation. Distal transformation of leg discs and of the wing disc is possible without an initial regeneration of all proximal structures around the circumference. Schubiger and Schubiger (1978), for instance, have found distal transformation in a fragment as shown in Fig. 9.3d without a preceding circumferential regeneration of the missing proximal structures. An analogous observation has been made by Karlsson (1980) for the wing disc. Our model is consistent with these experimentally discovered violations of the complete circle rule, since only a close juxtaposition of all major compartments is required.

The capability of a fragment which originated exclusively from the anterior

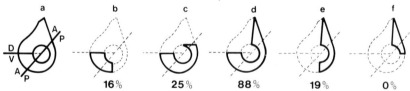

Fig. 9.3. Distal transformation in leg disc fragments. It occurs frequently if all compartmental specifications are present. (a) Leg disc with compartment borders according to Steiner (1976). The distal elements (tarsi and claws) are determined in the central part of the disc. (b–f) Frequency of distal regeneration of proximal leg fragments, according to Schubiger and Schubiger (1978) and Strub (1977a). The fragments b, c, e, f contain essentially only two compartmental specifications and show low frequency of distal transformation. In contrast, a fragment containing cells from all three compartments (d) frequently shows distal transformation in agreement with the proposed model (after Meinhardt, 1980).

leg compartment to show distal regeneration seems to contradict the model. However, compartment borders can be reformed during the regeneration of fragments. According to Schubiger and Schubiger (1978), the distal transformation of an anterior fragment is always associated with the regeneration of structures of the posterior compartment. The formation of new positional information for the proximo-distal dimension in such a fragment is assumed to be a two-step process. The first step is regeneration of parts of the missing compartment(s) (see Fig. 12.6). The second step is formation of a new morphogen distribution, centred over the new intersection of compartment boundaries. (In the polar coordinate model, ability of an anterior leg fragment to regenerate the missing members of the major compartments is accounted for by the assumption of a non-uniform spacing of positional values, see French et al., 1976.)

In the wing disc, the data as to what extent cells can change their compartmental specification after experimental interference are less clear (Garcia-Bellido and Nöthiger, 1976; Szabad et al., 1979). The A-P border seems to be more rigidly fixed than the D-V border. Thus it would be expected that a wing fragment, if it is to undergo distal regeneration, must include the antero-posterior compartment border. The dorso-ventral compartment border, since it can be respecified, would be of less importance. This is in agreement with the experimental observations of Karlsson (1980) and Wilcox and Smith (1980).

Small marginal fragments of a wing disc usually do not show distal transformation by themselves (Bryant, 1975a,b) because they contain at most cells of only two compartments. When two marginal wing fragments derived from opposite positions of the disc are joined together, they frequently show regeneration of the missing distal structures (Haynie and Bryant, 1976). This would be expected since these fragments together, generally contain cells from all major compartments. The same is valid also for an outer-ring fragment of a disc. Strong distal transformation also occurs after dissociation and reaggregation of imaginal discs (Strub, 1977b). This is expected from the model since in this procedure many new compartmental confrontations and hence morphogen sources are created.

The initiation of cell death in cell-autonomous cell-lethal mutants can lead to a partial or complete duplication or triplication of the leg (Postlethwait, 1978, Russell et al., 1977). According to the model, the primary event is a compartmental respecification. An explanation of how cell death can lead to a compartmental respecification is given below (see Fig. 12.7). It is especially common that structures belonging to the posterior compartment are formed in an anterior environment. Respecification can lead to new intersections of compartmental boundaries and therefore to the establishment of additional morphogen maxima (Fig. 9.4). An observation by Girton (1981) provides an

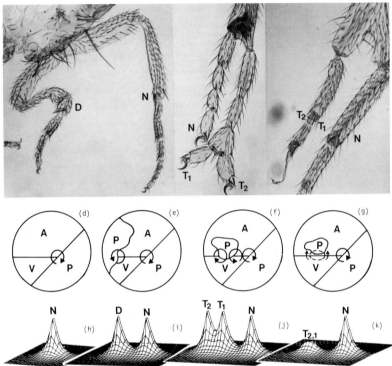

Fig. 9.4. Heatshock-induced leg duplications and triplications (Girton, 1981; Bryant and Girton, 1980). In addition to the normal leg (N) a single (D) or a pair of legs (T_1, T_2) is formed. In the latter case, the pair can be distally complete (b) or incomplete (c) and the pair can be partially (b) or completely fused (c) (photographs kindly supplied by J. Girton). Explanation in terms of the model: due to the heat shock (and cell death), part of the anterior compartment of a normal disc (d) becomes reprogrammed to posterior (see Fig. 12.7). This can lead to a new intersection (e) and consequent formation of an additional leg with opposite handedness (A, P, V clockwise or counterclockwise, see arrows). If the patch of posterior cells arises in a non-marginal position, two new intersections can be formed (f). This would lead to a pair of additional legs, as shown in (b). The closer the two new intersections are, the more distal the separation of the pair of legs will be. If the patch is close to but does not touch the ventral compartment border (g), cooperation is restricted, the maximum morphogen concentration is not reached and a fused pair of distally incomplete legs will be formed (as shown in c). Thus the model predicts that distally incomplete legs do not contain cells of the ventral compartment, in agreement with the observation (Fig. 9.5). The model provides an explanation of Bateson's rule (Bateson, 1880) according to which the three limbs are formed in a plane (the new intersections are formed along the AD-AV border line) and the central limb has opposite handedness when compared with the two others (see arrows). (h–k) Computer calculations of the positional information created by intersections shown in (d–g).

especially convincing example of the connection between leg duplications
and compartments. He found duplication and distally complete and
incomplete triplications of legs (Fig. 9.4). Drawing the structures of the
triplicated legs at the level of bifurcation in the fate map reveals that distally
complete outgrowth occurs only if cells of the ventral compartment are
present (Fig. 9.5). All duplicated legs contain the A-P boundary. Figures
9.4d-k show the expected locations of compartmental respecification and the
resulting morphogen distributions.

Some of the observed duplications and triplications indicate clearly that no
lateral inhibition is involved in this formation of new organizing regions.
Two interactions can appear so close to each other that the resulting legs are
fused over almost their entire length (Fig. 9.4b). No indication can be found
for competition or dominance of one leg over the other (in contrast, for
instance, to the formation of new heads in hydra, Fig. 6.2). Lateral inhibition
is the antagonistic reaction necessary to localize autocatalysis and to
suppress the formation of identical structures in the surroundings. If an
organizing region is formed by intersection of compartments, lateral
inhibition is not required since the intersection is confined *per se* to a
particular location.

In the abdomen, which consists of structures lacking a proximo-distal
dimension, no compartments have yet been found (Lawrence *et al.*, 1978).

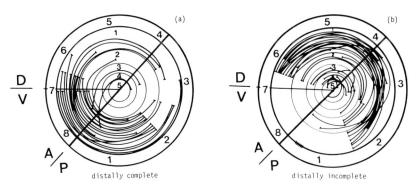

Fig. 9.5. Evidence that presence or absence of the ventral compartment is decisive as
to whether distally complete transformations occur. Shown are the fate maps of
distally complete (a) and incomplete (b) tarsal triplications. Each curved line indicates
the structures present at the base of a particular triplication (after Bryant and Girton,
1980). Distally complete legs contain cells of the ventral (V) compartment (a) while
distally incomplete legs do not (b). In agreement with the model, an antero-posterior
boundary is present in both types of legs. 1–8: tarsal bristle rows. The concentric rings
1–5: the five tarsal segments; thick black bars: approximate location of the two major
compartment borders according to Girton and Russell (1981). Note that it is not the
completeness of the circumferential structures (as proposed by Bryant *et al.*, 1981) but
the presence of all compartments which is decisive for distal transformation.

According to the proposed view, subdivision into compartments is a pre-condition for the generation of structures with a proximo-distal axis, for instance of wings and legs. Therefore it is not required in the abdomen. It could well be, however, that temporary A-P subdivisions also occur in the abdominal segments during the formation of segments (see Chapter 14).

Evidence for a morphogen gradient and a stepwise unidirectional determination

Distal fragments of a disc do not regenerate proximal structures (Schubiger, 1971; van der Meer and Ouweneel, 1974). Distal to proximal segmental respecification does not occur even if proximal and distal fragments are confronted (Haynie and Schubiger, 1979; Strub, 1979). This is in agreement with the proposed model since distal determination, once obtained, is assumed to be irreversible. (This is in sharp contrast to the distal-proximal regeneration within, for example, a leg segment (see Fig. 13.1) and emphasizes once more that different mechanisms are involved in these two types of pattern formation.)

Mutations are to be expected in which positional information is changed, but not the response of cells to it. Such mutations should not be cell autonomous. For instance, a small clone of mutant cells in a wild type environment is expected to develop like the wild type cells since they are exposed to the normal positional information. Two known mutations are of this type. *Drosophila* flies carrying the mutation *wingless* can duplicate the dorsal thorax but fail to form a wing blade. However, clones of *wingless* cells can and do participate in wing formation (Morata and Lawrence, 1977). According to the model, either compartmentalization or production of the morphogen by the cooperation of compartments may be affected. This mutant further suggests that the system which generates positional inform-ation for the wing is different from that of the leg since formation of the leg is not affected.

Jürgens and Gateff (1979) have found duplication of legs in a temperature-sensitive mutant (*mad*) of *Drosophila*. The orientation of the additional legs indicates that in this mutant a second dorsal compartment is formed at the ventral side of the disc while the antero-posterior axis is not affected. Mosaic studies have revealed that both mutant and wild type cells participate in the formation of the duplicated leg. This implies that the positional information and not the response to an unaltered gradient is what is changed. Distally complete duplications can be induced by a pulse of high temperature, applied between 48h and 76h after egg deposition. As the rule, the later the pulse, the more proximal the point of bifurcation. In terms of the model, the disc is larger (or subdivided into more cells) at a later stage and the intersections can, therefore, have a greater distance from each other. This leads to less overlap between the two systems of positional information and therefore to a more complete separation of the two legs.

Distally incomplete structures can occur in triplication of legs (mentioned above) as well as in cockroach legs after an injury (see Fig. 9.9). Since only the local concentration of the morphogen is interpreted, distally incomplete structures are expected if the normal maximum concentration is not reached. This can be caused by restricted collaboration of the compartments, e.g. if too few cells of a particular compartmental specification are available or if they are not in close enough proximity. The missing ventral compartment in distally incomplete leg triplication (Figs 9.4 and 9.5) directly supports this view.

Expected mutations

The model predicts mutations in which proximo-distal pattern formation is affected. The resulting pattern may be distally incomplete. In extreme cases, the whole disc may be missing. As mentioned, a mutation in which the generation of the signal (morphogen synthesis) but not the local response of the cells is affected would not be cell autonomous. This means that a clone of mutated cells in a wild type environment will participate in normal pattern formation. However, if such a clone arises close to the intersection of compartments, the prediction is that pattern formation in the whole disc is altered in that only distally incomplete structures are formed, even by the wild type cells. Further, the model predicts that several mutations exist with the same phenotype but that each mutation is specific for a particular compartment since the mutation effects a compartment-specific function in the cooperation. For instance, a clone has to arise in the posterior compartment and close to the intersection if a pattern alteration is to occur. Such genetic studies combined with DNA sequencing methods may allow one to trace the regulatory pathway of the morphogen responsible for proximo-distal determination. Distal transformation of posterior leg fragments (which normally do not show distal transformation, see Fig. 9.3f) could provide a bioassay for a putative morphogen.

Strategy for isolation of the morphogen

In a normal disc, the morphogen is produced presumably only in minute amounts rendering a biochemical characterization difficult. However, according to the model, a disaggregation and reaggregation of whole discs should lead to a tremendously increased morphogen production since cells of all compartmental specifications become close to each other at many locations, not only at the natural intersection as in the intact disc. Comparison of electrophoretic patterns of normal and of reaggregated disc may reveal spots of changed intensity, pointing towards the morphogen. Some experimental evidence is already available that this strategy may be successful. Reaggregates of dissociated leg discs show distal transformation extremely frequently (Strub, 1977b).

Application to pattern regulation in insect legs

In hemimetabolous insects, the adult appendage emerges not in a unique metamorphosis from a disc, but through a sequence of several moults. Between the moults, substantial pattern regulation is possible. The leg of cockroaches is a well-studied developmental system of these insects. Nothing is known about compartments in the cockroach leg, but as a working hypothesis we will assume a compartmentalization analogous to *Drosophila*. The compartments would have a stripe-like shape along the tube-shaped ectoderm of the leg. Many features of pattern regulation can then be made understandable by the proposed cooperation of compartments.

The cockroach leg, if removed, is capable of complete regeneration. During closure of the wound, cells of all three compartments come into close contact and positional information is regenerated. All those cells which are exposed to a higher morphogen concentration than that corresponding to their own specification undergo distal transformation. The regeneration of leg segments is assumed to be a reprogramming of existing cells and is therefore a morphallactic process. This view is supported by the observation of Bulliere (1972) showing that during leg regeneration the reformed segments are already clearly distinguishable by the time cell division starts (Fig. 9.6).

One strange result of these experiments is that, although a stump regenerates a complete leg, regrafting a leg fragment onto a cut stump can prevent regeneration of structures present neither in the stump nor in the leg fragment. If, for instance, a leg fragment consisting of a leg from mid tibia on is grafted onto a stump cut off in mid femur, the distal femur and proximal tibia will not be regenerated (Fig. 9.7). As explained in detail in Fig. 9.7, as growth occurs, positional information in most of the cells becomes lower than it was at the time of cell determination. After removal of an intermediate section of the leg, reprogramming is, as a rule, impossible. This failure of gap repair is typical for systems controlled by unidirectional interpretation of positional information, since the cells change their determination only when exposed to a higher morphogen concentration, not when confronted with an unnatural neighbour (see also Fig. 8.3). If, however, in such a regraft operation, wound healing is not perfect, two additional distal structures are

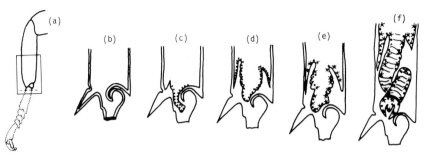

Fig. 9.6. Evidence that regeneration of an insect leg is a morphallactic process. (a) The location of the cut at the end of the tibia. The rectangle indicates the area shown in b–f. (b–f) Stages in the regeneration according to Bullière (1972). DNA-synthesis is marked by dots, cell division by an asterisk. Note that recognizable segmentation takes place *before* cell division starts (d). The initial formation of the new structure proceeds by respecification of existing cells and not by the building of new structures by proliferating cells.

generated (Fig. 9.7c). In terms of the model, as the ectodermal tube closes, cells of all compartments come into close contact and new centres of morphogen production are formed at both wound surfaces. This will lead to the reprogramming of the leg both proximal and distal to the wound such that symmetrical distal structures will be generated at the wound site (Fig. 9.7k).

After amputating a limb and reimplanting it either in a rotated position or onto a contralateral stump, supernumerary legs are formed (Bart, 1971a,b; Bohn, 1972; French, 1976b). On the basis of his experiments, Bart (1971a) has already proposed that new morphogenetic centres arise whenever different sides (anterior-posterior, dorsal-ventral) meet. Such an operation creates new intersections between compartments (Fig. 9.8). Their numbers and the handedness of the additional limbs will be discussed in detail for the amphibian limb system (see Fig. 10.6). Cutting a V-shaped notch into the ventral (internal) side of a leg leads preferentially to outgrowth of a symmetrical leg which is distally more or less complete (Fig. 9.9). This outgrowth is very striking if one expects only an intercalation between mismatching neighbours on the shortest possible route. A similar injury at the dorsal side heals with little, if any, outgrowth. Similarly, artificially produced double ventral legs regenerate to a large extent while double dorsal legs do not (Bohn, 1965). The model predicts this asymmetric behaviour. The explanation is given in Fig. 9.9.

In conclusion, very different and seemingly unrelated observations can be explained under the assumption that the borders between compartments are used to create new coordinate systems for the finer subdivisions of the

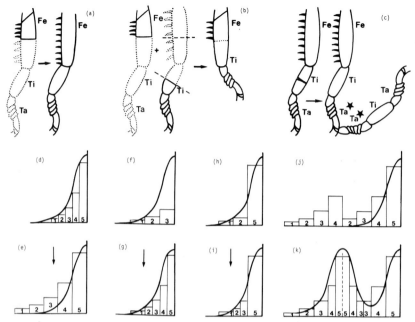

Fig. 9.7. Occurrence and failure of intercalary regeneration in systems controlled by positional information. (a) An amputated cockroach leg regenerates all missing parts. (b) However, grafting a mid-tibia (TI) onto a mid-femur (FE) stump does not lead to intercalary regeneration. The parts between the dashed lines remain missing (Bohn, 1970a). Thus, regrafting distal structures suppresses the formation of intervening parts. (c) Incomplete wound healing after cutting and regrafting can lead to distal transformation on both sites of the wound. Then, two additional tarsi (TA*) are formed (French, 1976a) analogous to the additional posterior structures in an *Euscelis* egg (Fig. 8.2e). (d–k) Explanation in terms of the model: it is assumed that the morphogen gradient (positional information) causes the determination of the structures 1–5. (e) During normal growth, the local morphogen concentration decreases in most of the cells; the gradient is assumed to be unaffected by growth. Since interpretation proceeds unidirectionally, the cells remain stable in their respective states of differentiation. (f) After removal of distal parts and reformation of the positional information, most cells are exposed to a higher morphogen concentration and all structures (g) are formed. (h) If an intermediate section of a leg is removed, the positional information that could lead to respecification of the structures at the wound is present only in the very terminal structures of the leg. Therefore, in most of the cells the positional information is lower than the level they were exposed to when the pattern was first established. Whether missing parts are replaced therefore depends on the extent of growth and the range of the morphogen. In this example (i), part of structure 2 becomes respecified to form 3 while structure 4 remains missing since the cells exposed to the appropriate morphogen concentration are already determined to form structure 5. (j) Analogously, no repair of a gap introduced by grafting surplus structures occurs since the morphogen concentration remains too low at the location of the gap. (k) If, however, such an operation triggers a new system of positional information, as would happen if, for instance, cells of all major compartments have come close together, two new distal structures will be determined by respecification. This corresponds to the experimental observation shown in (c).

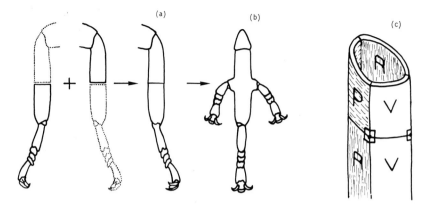

Fig. 9.8. Formation of supernumerary limbs after contralateral grafting of a cockroach leg (Bohn, 1965). (a) The operation. (b) Resulting pattern after several moults. (c) Explanation in terms of the model. Two new areas of confrontation of all three major compartments are created at the graft-host junction (squares), leading to two new limb fields.

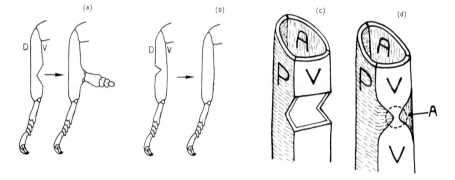

Fig. 9.9. Cutting a notch into the inner ventral (V) site of a cockroach leg (a) leads to outgrowth of additional leg-like structures. This result is very surprising if one assumes that juxtaposition of non-adjacent structures leads to an intercalation of missing structures only. (b) The same operation at the outer dorsal site (D) heals without much additional outgrowth (Bohn, 1965). (c) Explanation based on "cooperation of compartments". A subdivision similar to that of the leg of *Drosophila* into the anterior (A), posterior (P) and into the much smaller ventral (V) compartment is assumed. After removal of tissue at the ventral side, cells of anterior, posterior and ventral specifications become quite close to each other (d). Cooperation is possible and outgrowth of symmetrical distal structures is expected on the basis of the proposed model. Since anterior and posterior cells could be separated by some ventral cells, the cooperation may be restricted and distally incomplete structures can be formed. After a similar incision at the dorsal site, ventral cells remain far away and cooperation of compartments is impossible.

developing organism. A still hypothetical morphogenetic substance produced by the cooperation of compartments, provides positional information
about the distance of the cells from the border(s). Changes of the geometrical
arrangement of the compartments, caused either by surgical interference or
by a cell-internal switch in the compartmental specification can lead to new
intersections of borders and therefore to the formation of additional
structures.

10

Boundaries between differently determined cells control pattern formation in the limb of vertebrates

Polarizing and competent zones in the amphibian limbs

Cooperation of compartments has been suggested above as a straightforward mechanism to organized subfields in insects. In vertebrates, one of the best investigated systems of pattern formation in a developmental subfield is the limb (for review see Hinchliffe and Johnson, 1980). Grafting experiments reveal that cooperation of differently determined tissues is also involved in limb organization. In amphibians, two zones are important for limb development: the competent zone and the more posteriorly located polarizing zone (Harrison, 1921; Slack, 1976, 1977a,b). The future limb is formed almost exclusively from the competent zone. However, these competent cells can only form a limb when juxtaposed with cells of the polarizing zone. Grafting polarizing tissue anterior to the future limb area lead to the outgrowth of symmetrical limbs in which the posterior digits are duplicated (P-A-P pattern, Fig. 10.1). In some cases, two almost complete hands are formed while in others, some anterior digits are missing. From these experiments, Slack concluded that an interaction between the two zones is required to generate the positional information that controls anterior–posterior determination of the future limb. By implanting polarizing tissue from a salamander into an axolotl, Slack (1976) has provided direct evidence that only the competent tissue responds: the reduplicated leg consists entirely of axolotl type structures. After grafting tissue of the prospective limb area into a more posterior region of the flank, an additional limb with reversed polarity can result (Fig. 10.2). In terms of Slack's model, after such grafting, competent tissue is—in contrast to its normal position—located posterior to the polarizing zone leading to a reversed gradient and therefore to a reversed limb. It is easy to see how the two zones—the prerequisites for the limb

95

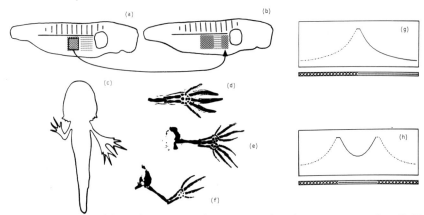

Fig. 10.1. Juxtaposition of two zones is necessary for the antero-posterior (A-P) organization of a limb. (a–c) Graft experiment by Slack (1976) with axolotls. (a) Donor embryo and the location of the polarizing (× ×) and competent zone (= =) as determined by this and by the experiment described in Fig. 10.2. (b) Host embryo after implantation of the polarizing tissue anterior to the competent zone. (c) The resulting symmetrical (right) limb. The posterior digits are always present and duplicated while some anterior digits may be missing. (d, e) Bones of symmetrical limbs and of a normal limb (f) (after Slack, 1977a,b). (g, h) Model: in the normal situation, confrontation of polarizing and competent tissue leads to a symmetrical morphogen distribution, centred over the common boundary. Only the competent cells can respond and this leads to a monotonic gradient (solid line). After a graft as shown in (a, b) the competent zone is confronted with polarizing tissue both at its anterior and posterior margins, leading to a symmetrical morphogen distribution (h) and therefore to a symmetrical arrangement of skeletal elements. Depending on the overlap of the two gradients, low concentrations and therefore anterior digits. could be absent.

formation—might be formed during development. Interpretation of a primary antero-posterior gradient in the embryo could lead to several belt-shaped patches of differently determined tissues. Two of these could be the polarizing and the competent zones.

Independent of the question of how a limb field is formed, these experiments provide an important indication concerning the origin of polarity in tissues of higher organisms. Harrison (1921) believed that overall polarity results from the superposition of many small polar structures, e.g. polar cells. In an experiment such as that shown in Fig. 10.2 the graft is only transposed, not rotated; the A-P orientation of the graft remains unchanged. Thus, the experimentally observed change in the polarity of the outgrowing limb was regarded as very striking. In fact, the experiment provides strong evidence that polarity does not result from many small polar substructures but from the slope of graded distributions of morphogenetic substances.

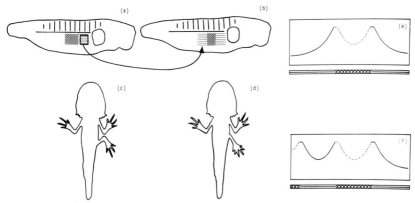

Fig. 10.2. Polarity reversal results from grafting a competent zone posterior to a polarizing zone. (a–d) Grafting experiment that demonstrates the existence of a competent zone (after Slack, 1976, 1977a,b). Tissue located ventrally to somite 3–5 is transplanted into a more posterior position (a, b). This leads to outgrowth of an additional leg which either has reversed A-P polarity (c) or is of the symmetrical P-P-type (d). (e, f) Model: after the grafting operation, the grafted competent zone is located posterior to a polarizing zone, so that a gradient with reversed polarity is formed (e). If some of the polarizing zone of the host is included in the graft, a symmetrical pattern is formed (f).

Generation of polar structures by cooperative interaction between two differently determined patches of cells

The need for cooperation between the two zones is reminiscent of the mechanism proposed above for pattern formation in imaginal discs. However, a major difference remains. The vertebrate limb is a structure with clear antero-posterior polarity while the fate map of the leg disc of *Drosophila* shows that the elements are circularly arranged (see Fig. 9.2). As has been shown above, cooperation of two zones leads to a symmetrical morphogen distribution, centred over the common boundary. Therefore each zone contains one of the two slopes of the morphogen ridge. If only one of the two zones is able to respond, the cells of this zone are exposed to an exponential gradient. The end result is a polar instead of a symmetric structure (Fig. 10.1).

Two intersecting boundaries are required to determine a limb field

The common border of the polarizing and competent zone would have a long extension in the dorso-ventral direction. The morphogen distribution in the

competent cells resulting from the cooperative interaction is high along the
whole posterior boundary of the competent zone and low at its anterior side.
This morphogen distribution is therefore only able to organize the antero-
posterior axis of the limb. The dorso-ventral position of the outgrowing limb
remains to be determined. As in imaginal discs, this can be achieved by an
intersection with a second boundary (Fig. 10.3) which results from a pattern-
forming event organizing the dorso-ventral axis (see p. 150). This intersection
of two borders completely defines the limb field. The intersection itself
determines the position of outgrowth of the proximo-distal axis. The distance
from the competent-polarizing border determines antero-posterior position.
The dorso-ventral (D-V) pattern is more or less symmetrical. It is to be
expected that both slopes of the resulting symmetrical D-V morphogen
distribution are used but that they are differently interpreted in the dorsal
and ventral parts. Since the intersections depend on the primary organization
of the embryo, not only is positional information within the limb field
generated but the limb field necessarily has correct orientation with respect to
the body axis and correct handedness. (Pattern formation in the proximo-
distal axis will be dealt with in detail on p. 117. A mechanism of
interpretation of the A-P gradient leading to superposition of sequential and
periodic structures as indicated by the sequence of digits, will be discussed in
Chapter 14.)

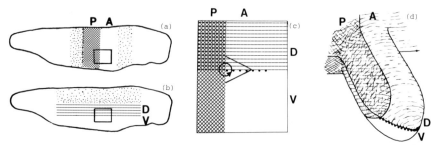

Fig. 10.3. Steps in the formation of the limb field. (a) A primary antero-posterior
gradient can serve to subdivide an embryo into bands of distinct determinations (see
Figs 11.5 or 14.5). Among these are the polarizing (posterior, P, crossed) and the
competent (anterior, A, blank) zones. (b) To locate the position of the outgrowing
limb, a global dorso-ventral subdivision of the embryo is required. This leads to the D
(= =) and V (blank) stripes. The area around the intersection (○) of the A-P and D-
V border (framed in a and b; c) defines the position of the limb field. Cooperation of
the A and P and the D and V tissue lead to a symmetrical AP and DV-morphogen
distribution. Since only the competent cells respond, the antero-posterior pattern is
polar (see Fig. 10.1). The positional information generated in this way (indicated by
the triangle) is a measure for the distance of a cell from the borders. The dotted line
marks the expected position of the apical ectodermal ridge (AER) on the D-V border
in A (competent) tissue. (d) Geometry of the A-P and D-V stripes in an outgrowing
right limb bud, viewed from a posterior-dorsal position.

Regeneration and formation of new limb fields after experimental manipulations

It is not yet possible to test such a model at the biochemical level. Support is provided by the demonstration that the model correctly predicts the altered pattern resulting from certain experimental interferences.

Some amphibians regenerate amputated distal parts of a limb or form supernumerary limbs after other experimental interferences. According to the model, a new limb field is formed if experimental interference leads to a new intersection of both boundaries. Under this assumption, regeneration of a limb indicates that the limb does not consist entirely of competent tissue but that at least a small stripe of polarizing tissue is carried along with the outgrowing limb. After truncation of a limb and closure of the wound, a new intersection can be formed, which enables reformation of the removed parts.

Presence and absence of regeneration of experimentally produced symmetrical limbs

A critical test of any model for limb development is whether it can account for the very striking differences in the regeneration capability of different types of symmetrical limbs. A symmetrical limb consisting of two posterior halves (PP-limb) regenerates a PP leg (Slack and Savage, 1978a,b). In contrast, a limb consisting of two anterior halves (AA) shows little, if any, regeneration (Stocum, 1978). These results are easy to understand in light of the proposed mechanism. Clearly, a PP-limb has two stripes of polarizing tissue, one on each site (Figs 10.1 and 10.4). This leads to two intersections of all boundaries and hence to a symmetrical P-A-P pattern with some of the most anterior structures missing. After removal of distal parts of such a PP leg, two intersections of the two boundaries are reformed and therefore the original symmetrical P-A-P pattern is re-established. The formation of the second posterior side is, if the two intersections are sufficiently separated, connected with the formation of a second parallely aligned proximo-distal axis. This shows the interdependence of the two axes.

The failure of experimentally produced double anterior (AA) half limbs to regenerate is also described correctly by the cooperation model. An AA limb contains no P strip on either side (Fig. 10.4). Only a D-V border is present which is neither sufficient to trigger a proximo-distal outgrowth nor an A-P organization. The different behaviour of AA and PP limbs results from the asymmetry of the leg, which contains mainly competent (A) and only a small stripe of polarizing (P) tissue although both tissue types have to be present if distal outgrowth is to occur. Other observations on the regeneration of AA or PP legs are less well understood. The frequency of regeneration depends

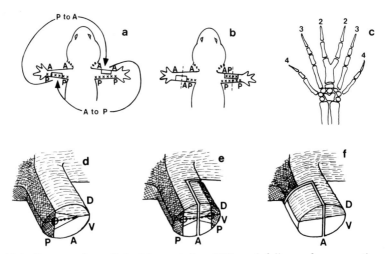

Fig. 10.4. Regeneration of double posterior (PP) and failure of regeneration of double anterior forelimbs. (a, b) Surgical production of symmetrical PP and AA limbs by reciprocal exchange of anterior and posterior halves of the forelimbs of axolotl with subsequent removal of the distal parts (b). (c) Result: symmetrical limb regenerated by a PP stump. The most anterior digit 1 is missing, 2–4 are duplicated (after Holder *et al.*, 1980). An AA stump shows little or no regeneration. (d–f) Wound closure in a normal, in a PP- or in an AA-limb leads to one (d), two (e) or no (f) intersections (circles) of the two borders. Either a normal, a PP leg (as in c) or no leg is expected to regenerate. The triangles indicate schematically the A-P and D-V morphogen profiles, the dots indicate the position of the AER.

critically on the time between the surgical production of a symmetrical limb and the removal of its distal part (which eventually induces regeneration). The shorter this time is, the higher the frequency of regeneration (Stocum, 1978; Tank and Holder, 1978; Bryant and Baca, 1978). In contrast to upper arm AA-stumps, lower arm or leg AA-stumps regenerate almost as well as PP-stumps (Stocum, 1978; Krasner and Bryant, 1980).

Formation of supernumerary limbs after rotation or contra-lateral grafting experiments

It has been known for many years (Bryant and Iten, 1976), that cutting a limb blastema and regrafting it after 180° rotation can lead to outgrowth of supernumerary limbs. The same happens after grafting a limb tip onto a contralateral stump. In Fig. 10.5, examples of the observed results are shown. It is difficult to imagine offhand which and how many intersections are formed by a particular operation. In Fig. 10.6, the leg cylinder has been

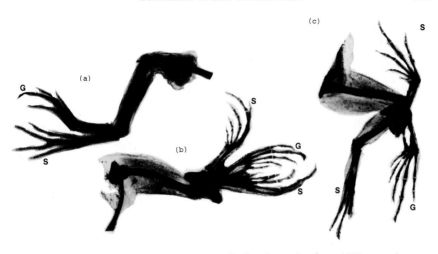

Fig. 10.5. Examples of supernumerary limbs formed after 180° rotation or contralateral grafting of a limb blastema in *Rana temporalis* (Maden, 1981b). (a) Result after a 180°-rotation: a symmetrical (PP) limb is formed in addition to the graft (G). (b) 180°: two supernumeraries (S) with normal antero-posterior polarity are formed in addition to the graft (G). (c) Contralateral graft with D-V inversion: two "normal" supernumeraries. (Photographs kindly provided by Malcom Maden.)

unrolled and the location of the borders at the graft–host junctions are shown. It can be seen that in all these operations, at least two new intersections of the two borders are created, enabling formation of two supernumerary limbs. Figure 10.6 shows further that the model predicts a very striking difference between a 180° rotation and a contralateral graft of a limb tip. After 180°-rotations, very complex intersections occur which depend on the precise alignment between host and graft. Especially, only after 180° rotation formation of symmetrical (PP) limbs is possible. Such complex supernumeraries have been observed by Maden (1980, 1981, 1982). Figure 10.5a shows an example, in which the symmetrical PP-limb can be recognized easily by its bifurcated central digit. The situation is very different after transplantation of a limb blastema to the contralateral side. Depending on the graft, either the A-P or the D-V axis is inverted. According to the model, this leads to two normal but never to PP-type intersections (Fig. 10.6). These predictions are fully supported by Maden's experimental observation. The model leads to an even more precise prediction: an A-P confrontation should lead to one supernumerary limb derived from the host, the other from the graft. In contrast, after a D-V confrontation, host and stump tissue should contribute to each supernumerary limb. Either the dorsal side is derived from the host and the ventral site from the graft or vice

versa (Fig. 10.6d,c). Whether this prediction is true has to wait for further
experimental investigations.

The experiments of Maden (1981) show further a substantial variability in
the frequency of supernumerary outgrowth between different amphibian
species. Even if the same operation is made repetitively on the same species,
the resulting pattern cannot be predicted with certainty. Only a probability
for the formation of one or two supernumeraries can be given. The intersection
of the boundaries appears therefore to be a prerequisite for a distal outgrowth
but other factors such as blood supply or innervation must be involved in the
decision whether it occurs or not.

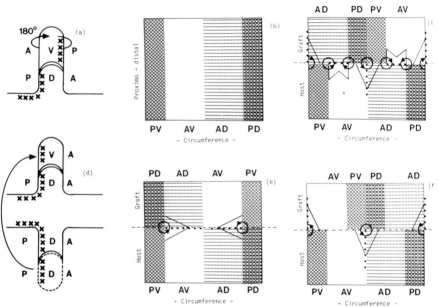

Fig. 10.6. Explanation of supernumerary limb formation. (a) After 180° rotation, A
and P as well as D and V tissue become juxtaposed at the host-graft junction.
(b, c) To visualize the arrangement of tissue with A (blank), P (×××), D (---)
and V (blank) specificity, the leg cylinder (see Fig. 10.3d) has been unrolled. Cells at
the right and left margin of the figure are, in reality, neighbours. Condition for a limb
field is a DV border in an A-area (dashed/blank border), flanked by posterior
(crossed) tissue. As in Fig. 10.3, a limb field is indicated by circles, triangles and dots.
(b) In a non-operated limb, the A, P, V and D specificities are arranged in stripes,
oriented proximo-distally. No intersections occur. (c) Arrangement after 180°
rotation of the graft (top): 6 intersections occur, two can lead to normal limbs
(triangles). The other 4 can lead to symmetrical double posterior (PP) legs, indicated
by the M-shaped morphogen distribution. (d-f) *Contralateral* grafts lead either to a
DV (d, e) or an AP-juxtaposition (f). After both types of contralateral transplantation
only two normal intersections are formed (no PP-legs) and the supernumerary legs
have the handedness of the stump (arrows).

Pattern formation in the chicken limb bud

Many experimental data are available concerning the developmental control of the chicken limb bud, revealing both differences and similarities to that of the amphibian limb. The most obvious difference is that after truncation, an amphibian limb regenerates while a chicken wing bud does not, unless an apical ectodermal ridge (AER) is transplanted onto the stump. The AER is a thickening of the ectoderm on the wing bud oriented parallel to the antero-posterior axis of the embryo.

A second area important for the pattern formation in the bud is the "zone of polarizing activity" (ZPA), discovered by Gasseling and Saunders (1964). The ZPA is a nest of mesodermal cells located at the posterior margin of the bud. It does not contribute to the limb proper but instead it develops as the "posterior necrotic zone". Upon transplantation into a more anterior position, the ZPA can induce a symmetrical antero-posterior pattern and a second proximo-distal axis. The grafted ZPA establishes the posterior side of the additional limb structures analogously to the situation observed after transplantation of polarizing tissue in amphibians (Figs 10.1, 10.2). On the basis of its location and orientation, it is tempting to identify the AER with the D-V boundary as discussed above. This view is supported by the observation that in dissociation-reaggregation experiments, the ectodermal hull determines the dorso-ventral axis of the leg (MacCabe *et al.*, 1973, 1974). However, transplantation of early limb bud mesoderm under an etopic ectoderm can induce an AER in the ectoderm (Kieny, 1960), showing that the primary DV-organization also takes place in the mesoderm. The ZPA, controlling the antero-posterior axis, is, as mentioned, of mesodermal origin. The ZPA is only effective when transplanted close to an AER (Wolpert *et al.*, 1975). Therefore, similarly to the situation in amphibians, an intersection of an AP-border (the ZPA in the mesoderm) and a DV border (the AER in the ectoderm) seems to be required for generation of a limb field. AER and ZPA appear to be specialized tissues which can be formed only during an early embryonic stage. The signal for their formation at the appropriate location results presumably from such D-V and P-A juxtaposition. If a limb bud is truncated at a later stage, D and V-tissue again comes into contact during wound closure but at that time, an AER can no longer be induced and therefore, the truncated bud fails to regenerate.

Wolpert and his coworkers (Tickle *et al.*, 1975; Wolpert *et al.*, 1975; Summerbell, 1979) have shown that the pattern formed after ZPA transplantation can be explained under the assumption that the implanted ZPA acts as a local morphogen source (Fig. 10.7). Implantation of a ZPA for a limited period of time indicates that the response of the cell is completely analogous to that deduced from the insect experiments (see Fig. 8.7). The cells are "promoted" stepwise and unidirectionally towards a more posterior determination until the actual determination corresponds to the local morphogen concentration. The cells remain stable in an achieved state of determination after removal of the morphogen source (ZPA). This mode of interpretation resolves a long-lasting controversy (Saunders, 1977) about whether the ZPA is involved at all in normal pattern formation. It has been found experimentally that the ZPA can be removed relatively early without preventing normal development (Fallon and Crosby, 1975). The ZPA appears to be superfluous. But at the same time, the ZPA can induce

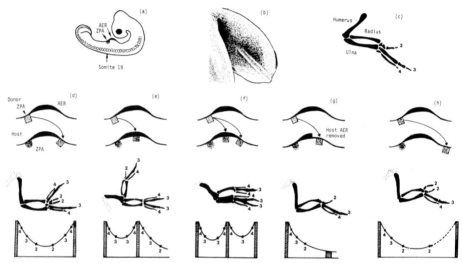

Fig. 10.7. Determination of the antero-posterior pattern in the chicken wing. (a) The chicken embryo with the wing bud between somite 16 and 19. (b) A wing bud at stage 23 (drawn after Hinchcliffe and Johnson, 1980) showing the apical ectodermal ridge (AER). (c) Normal pattern of a wing. (d–h) ZPA-graft experiments (based on Wolpert and Hornbruch, 1981; Tickle *et al.*, 1975; Summerbell, 1974a): the operations (top), the result (centre) and the explanation on the basis of a gradient model. (d) Graft of a ZPA to the anterior side leads to symmetrical development. The ZPA is assumed to be the source of the morphogen. (e) Implantation of the ZPA into the centre can lead to a complete limb pattern in the anterior part and an incomplete symmetrical pattern in the posterior part. (f) Two ZPAs grafted into the centre and into an anterior position can lead to the digit pattern 434 in the anterior part. This result argues strongly against an intercalation of structures since the anterior digit 2 is *not* formed. (g) For the action of the ZPA, close contact with the AER is required. After removal of the anterior part of the AER, implantation of a ZPA does not change the pattern. (h) Graft of a ZPA outside the proper limb field (opposite somite 15). Most of the cells exposed to the high morphogen concentration are incompetent, only digit 2 is duplicated, indicating that pattern formation depends on a long range signal and not on a chain of inductions between neighbouring cells.

additional structures as mentioned above. In terms of the model, the unidirectional interpretation of the antero-posterior gradient is completed relatively early. After removal of the source, the cells remain stable in the achieved state of determination. Nevertheless, after implantation into a more anterior position, the increased morphogen concentration is able to reprogramme the cells from their more anterior to a more posterior determination state. After very early excision of the ZPA, it is expected that the new juxtaposition of P and A tissues leads to regeneration of the ZPA and therefore also to normal development. Therefore, normal development can follow after ZPA removal in any case.

Tickle (1981) determined how many ZPA cells are required to induce additional digits. About 35 cells are sufficient to induce an additional digit 2 (the most anterior one, demanding the lowest morphogen concentration) and 100 cells can induce the complete sequence. These numbers are surprisingly small. However, if cooperation is involved, it is expected that only the marginal cells which are in contact with the AER contribute to morphogen production. If the number of cells is small, almost every cell has contact and contributes. Another important piece of information can be deduced from this result. Induction of the antero-posterior axis in the vertebrate limb is not a self-amplifying process infecting surrounding cells. It is not characterized by a clear threshold and an all-or-nothing result as observed in determination of the antero-posterior axis of an insect embryo (see Figs 8.1 and 8.4) or in the induction of new heads in hydra (Fig. 6.2). For generation of the primary embryonic gradient, we have had to assume autocatalytic mechanisms. In contrast, if the morphogen gradient were generated by cooperation of cells, a graded (not a switch-like) quantitative relationship between cooperating cells and morphogenetic level would be expected. This suggests a strategy for isolation of the morphogen. Creating a close contact between many AER and ZPA cells, for instance, by disaggregation and common tissue culture, should increase the morphogen production dramatically.

Relation to the polar coordinate model

The proposed mechanism provides a molecular basis for the complete circle rule proposed by French et al. (1976, see p. 81). They stipulate that distal outgrowth occurs whenever a complete set of 12 circumferential positional values are close to each other. Their choice of 12 values was somewhat arbitrary. According to a revised version (Bryant et al., 1981), distalization occurs locally whenever the positional values are complete in a particular area of the circumference. If three or four values were selected instead, the two models would have similarities, since demanding a complete set of four quadrants is equivalent to requring an intersection of two borders. Both models therefore make the same predictions about the handedness of supernumerary limbs since handedness is independent of whether 3, 4 or 12 positional values are assumed. Both models therefore predict that, after a contralateral graft, both supernumeraries should be of the stump handedness. Despite these similarities, the models make different predictions. According to the model presented above, the morphogen gradients are set up and distal outgrowth occurs whenever an intersection between an A-P and D-V border is present. In contrast to the assumption of the complete circle rule, distal outgrowth is assumed to be independent of the limb's fine structure which results from the interpretation of these gradients. The experiment mentioned above in which polarizing tissue is grafted into a more anterior position (Fig. 10.1), should illustrate the different predictions. According to the proposed cooperation model, two intersections of all borders appear close to each other. This leads to two parallel outgrowing proximo-distal axes and a symmetrical P-A-P pattern. In agreement with the experimental observation, outgrowth is expected despite the fact that some of the most anterior structures are missing due to overlap of the two resulting morphogen distributions. In contrast, the complete circle rule would predict that missing anterior structures would lead to an incomplete proximo-distal outgrowth at best. For ipsilateral 180°-rotation the proposed model predicts the observed complex supernumeraries which was unexpected on the basis of the polar coordinate model. Further, the model presented links the regulatory properties of developing appendages with primary pattern formation while, on the polar coordinate model, it remains an open question as to how the circumferential positional values are generated in the first place, how the handedness of a limb is determined and how a limb obtains its correct position and orientation with respect to the body axes. Nevertheless, the model of French et al. served to stimulate focused experimentation which contributed substantially to our present knowledge. It has narrowed the possible molecular mechanisms since most of the model's predictions have proved to be correct and it helped in developing the model discussed above.

Boundaries in other types of embryonal induction

The creation of a new coordinate system by "cooperation of compartments" is presumably not restricted to appendages. For instance, the diencephalon, the central structure of the forebrain, organizes the adjacent optic tectum (Chung and Cooke, 1975). Normally, the diencephalon is located anteriorly to the tectum. Operations in which diencephalic tissue becomes located at a posterior position lead to a polarity reversal of the tectum as visualized by a reversal of the retino-tectal connections. This result is completely analogous to the polarity reversal of limbs after grafting of competent and polarizing tissue (Figs 10.1 and 10.2).

Recent rotation experiments with the eye primordium of *Xenopus* have revealed that the eye has a stable antero-posterior polarity even at the earliest stage in which the rudiment can be detected (Gaze *et al.*, 1979). The eye results from an inductive interaction between an outgrowing part of the forebrain and the ectoderm. Extending the model developed above, the position of the outgrowth is presumably determined by a hidden boundary (as is the case in insect legs). The determination of "anterior" and "posterior" precedes that of the eye itself. Therefore, a non-polarized eye cannot exist, just as an imaginal disc cannot be formed without a preceding subdivision into compartments.

The same mechanism—cooperation of "compartments"—is used in such distantly related organisms such as insects and vertebrates to organize the substructures. It is, therefore, a very general mechanism. It assures that the newly formed structures have the correct position and orientation in relation to the parts already existent in the developing embryo.

11
The activation and maintenance of determined states

The geometry of a morphogenetic field is quite restricted if pattern formation mechanisms are involved which depend critically on diffusion. Since the time required for communication amongst all the cells of a field increases quadratically with the mean dimension of the field, pattern formation must take place within small assemblies of cells. As has been pointed out by Wolpert (1969, 1971), embryonic fields are indeed small, of the order of 1 mm or 100 cells across, a size in which communication via diffusion can take place in a few hours. As this initially small field grows to attain its final size, it becomes necessary that either the pattern-forming mechanism or the response of the cells is turned off. Otherwise, as the field of cells enlarges, periodic prepatterns may emerge, which would destroy the normal relationship between body parts. By the instructions a cell obtains during its early developmental history the cell becomes determined for a particular developmental pathway. Determination would be a long-term memory for the developmental signals to which the cell was exposed. This memory should be maintained even when the cells come into contact with cells of different determination, for instance, due to folding of cell sheets or due to migration of individual cells (such as neural crest cells) through the organism.

Biochemical switches

The maintenance of a particular determined state is a dynamic process. For instance, imaginal disc cells can maintain their state of determination over many generations in tissue culture. However, transdetermination—abrupt changes in determination—can occur (Hadorn, 1967; Gehring and Nöthiger, 1973), indicating that determination involves stabilization of a particular state and suppression of alternative pathways. A particular determined state is presumably characterized by activation of a characteristic set of genes. Such a state can be maintained by feedback of the activated genes upon their own activity. This would occur if a gene is transcribed in the nucleus, the associated mRNA is then transported into the cytoplasm and directs there the synthesis of a protein which activates the further transcription of that gene. Non-linear feedback-loops enable the formation of two or more stable states which can be selected by external signals and in which the cell would remain even after a removal of these signals.

With appropriate parameters the following simple reaction has two stable states (Meinhardt, 1976; Lewis *et al.*, 1977):

$$\frac{dg}{dt} = \frac{g^2}{1 + fg^2} - eg + m \tag{11.1}$$

The substance g has a non-linear, saturating feedback on its own production and a normal first-order decay. At low g-concentration, the negative linear decay term dominates, dg/dt is negative, and the g-concentration decreased further (Fig. 11.1). At higher g-concentration, the quadratic production term becomes important and the concentration increases until saturation is reached. A well-defined threshold exists and only two stable states are possible (Fig. 11.1).

An externally supplied morphogen, m, can cause a transition from one

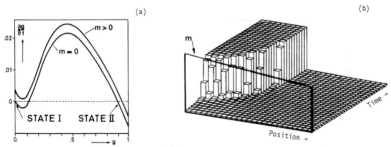

(a) (b)

Fig. 11.1. A model for the switching behaviour (determination) of a cell. (a) Only two stable states (I and II) are possible if a substance, g, has an autocatalytic, saturating feedback on its own production and is broken down by a normal first order process (eq. 11.1). If the g-concentration is above a threshold, the change per time unit (dg/dt) is positive. Then, the g-concentration increases until the saturation (state II) is reached. An additional g-production under the influence of a morphogen can lead to positive dg/dt-values also at low g-concentrations. The state I is no longer stable and the system switches irreversibly into state II (Meinhardt, 1976). (b) A shallow gradient of the morphogen (m) is assumed and the g-concentration is plotted as function of position and time. Only those cells exposed to a concentration above a threshold, switch to the high g-concentration. Despite the shallowness of the signal, the cells respond unambiguously (calculated with eq. 11.1, $e = 0{\cdot}1, f = 10$).

such stable state to the other. If m contributes to the production of g, the g-concentration can be pushed above the threshold and the cells made to switch irreversibly to the state of high g-concentration. Even a shallow gradient of the morphogen distribution can cause an unequivocal separation of an area into distinct sub-areas with cells of high and low g concentration, respectively (Fig. 11.1b). Alternatively, developmental decisions could be characterized by a restriction of possible pathways, involving the switching off of sets of genes rather than the activation of additional genes. This is possible if the morphogen interferes with the feedback, bringing the g-concentration temporarily below the threshold. In both cases, the cells would remain in the new state even after withdrawal of the morphogen.

Alternative states

Determination can also consist of the selection of alternative pathways (in contrast to the possible activation of one feedback loop). For instance, either gene 1 or gene 2 becomes activated but the cell must decide between these alternatives since both gene activities are mutually exclusive within one cell. A simple interaction with such a switching behaviour is described by the following equations:

$$\frac{dg_1}{dt} = \frac{1}{a + g_2} - g_1 \qquad (11.2a)$$

$$\frac{dg_2}{dt} = \frac{1}{b + g_1^2} - g_2 \qquad (11.2b)$$

(Production and decay rates have been set arbitrarily to unity, a and b are introduced to give a Michaelis–Menten cinetics.) It is easy to see that the switching behaviour of eq. 11.2 has a similar formal basis as that of eq. 11.1. Calculating the steady state of g_2 ($dg_2/dt = 0$) and neglecting b, we find $g_2 = 1/g_1^2$. Inserting this into eq. 11.2a leads to:

$$\frac{dg_1}{dt} = \frac{g_1^2}{1 + ag_1^2} - g_1,$$

which is identical with eq. 11.1.

A biological system in which the selection of two alternative pathways depends on the mutual repression of two genes is the Lambda phage. The DNA of the phage can either be integrated into the E. coli chromosome and replicate accordingly (lysogenic mode) or can replicate independently and thereby eventually kill the host cell (lytic mode). A simplified reaction scheme is given in Fig. 11.2. This reaction has an even higher non-linearity than given

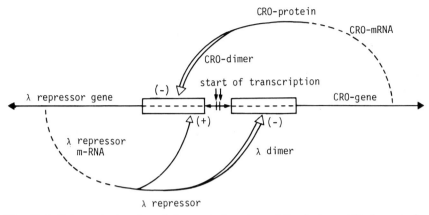

Fig. 11.2. Schematic drawing of the control system of the λ-phage (Ptashne *et al.*, 1980). Transcription of the λ-repressor gene leads to the repression of the CRO-gene and vice versa. Only two stable states are possible, the lytic (CRO on, λ off) or the lysogenic (λ on, CRO off).

in eq. 11.2, since both repressing substances are active as dimers and, in addition, the λ-repressor has an autocatalytic feedback on its own mRNA transcription (Ptashne *et al.*, 1980).

Similarities between pattern formation and the selective activation of genes

For many developmental decisions, presumably more than two alternatives must be envisaged. How can one stabilize one out of, let us say, ten genes and suppress all the others? The selective activation of genes has many formal similarities with the formation of a pattern. In pattern formation, only a few cells (those in a given region) become activated, the others are inhibited. Similarly during determination, only a few particular genes become activated, the others are repressed. Determination is, so to speak, pattern formation in gene space and it is tempting to assume formally similar mechanisms.

An interaction analogous to eq. 3.1 for pattern formation is given in eq. 11.3:

$$\frac{dg_i}{dt} = \frac{c_i g_i^2}{r} - \alpha g_i \tag{11.3a}$$

$$\frac{dr}{dt} = \sum_i c_i g_i^2 - \beta r. \tag{11.3b}$$

Each gene i ($i = 1, 2 \ldots n$) of the set receives feedback from its own activity via a gene activator g_i (Fig. 11.3a). This has the consequence that a gene, once activated, remains activated. An antagonistic reaction is required in addition, otherwise every gene would be activated. This can be brought about by a repressor, r, which is produced by every active gene and which acts upon every gene belonging to the set of alternative genes (Fig. 11.3b). We have seen that differing diffusion rates play an important role in pattern formation. The corresponding parameter in the selective activation of genes is the specificity of the activating and inhibiting molecules. The low diffusion rate of the activator molecules corresponds to a weak cross-reaction of the different g_i molecules with one another. The autocatalysis is either position-specific as in pattern formation or gene-specific as in gene activation. In contrast, the inhibitor acts, due to its redistribution by diffusion, upon every cell of a field. Correspondingly, the repressor has to act upon every gene belonging to the set of alternative genes. In such a system, two active genes of the same set within one cell would create an unstable situation. Each would compete with the other via the common repressor (Fig. 11.3b): the dominating gene attains a stable equilibrium with the repressor. A decrease of a g_i concentration, for instance, would lead to an overproportional reduction in the r concentration, enabling a readjustment. Instead of utilizing a repressor, the mutual exclusion of the (autocatalytic) genes can also be obtained by competition for a common precursor molecule, in analogy to the pattern formation eq. 5.1.

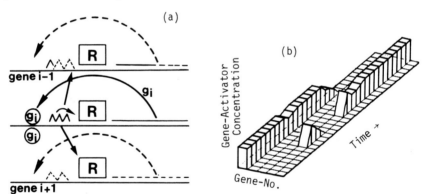

Fig. 11.3. Alternative stable states. Determination requires that a cell can remain stable in one of several alternative states. (a) A molecular interaction on the level of the genes which can accomplish this. Each of the alternative genes i ($i = 1, 2, \ldots$) feeds back on its own activation via a gene activator g_i; but competes with the others via the common repressor R ($\wedge\!\!\wedge\!\!\wedge$ = repressor coding site). (b) Computer-simulation (eq. 11.3). Only one of the feedback loops (gene 1, 2 ...) is stable within a given cell. Artificial activation of another loop will either decay or win the competition, suppressing the previously active gene.

Taken together, the simple reaction described in eq. 11.3 has two essential properties: a gene, once activated, remains active; and activation of different genes is mutually exclusive; only one of the several alternative genes can be active within one cell. The formalism developed is general and can be applied to many self-stabilizing systems. The activation of particular genes is only one of the more straightforward interpretations. The major question that remains is how a system chooses the correct state from several alternatives. As mentioned, experimental evidence suggests that there are two possibilities: either mutual induction of such locally exclusive states (p. 138), or control by a morphogen gradient (see below).

Interpretation of positional information

Evidence has been provided that the local concentration of a substance, the morphogen, is used to select a particular developmental pathway. The graded distribution of the morphogen controls, therefore, the formation of an ordered sequence of structures. The question then is how this positional information is to be interpreted. How to convert the labile morphogen concentration, which would be sensitive to any change in the geometry of the

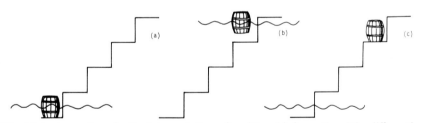

Fig. 11.4. An analogy for the interpretation of positional information. The differently determined stable states which a cell can occupy are compared with the different levels on which a wooden barrel can rest on a staircase. Only a few alternatives are possible. The local morphogen concentration corresponds to a flood, lifting up the barrel. After the flood has gone, the level at which the barrel comes to rest is a measure of the height of the flood. The lifting up is essentially irreversible. A second, higher flood can lift the barrel to higher levels while its withdrawal is without effect. The flood analogy can be used to underline the differences between two expressions frequently used in this context and give them a clear definition: positional information and positional values. Positional information denotes the level of the external signal, the height of the flood. In contrast, the positional value would describe the result of the interpretation, the status of the system, the level the barrel has attained. The two parameters have different properties. Positional information requires cell communication. The isolation of a small group of cells from the surrounding tissue will lead, as the rule, to a change in the positional information while the positional value is a stable property of the state of determination and would remain unchanged after withdrawal of the positional information.

developmental field, into a stable state of determination. A biochemical analogue-to-digital converter is required. There are several ways that the local concentration of a morphogen can serve to activate genes selectively. From ligation experiments with insect embryos, it has been deduced that the cells do not measure the local concentration all at once but rather that they are "promoted" step by step, switching from a more anterior (or proximal) to more posterior (distal) state until the state achieved corresponds to the local morphogen concentration (see Fig. 8.7). This narrows down the possibilities of interpreting positional information considerably. A straightforward mechanism consisting of the direct sequential activation of the genes until the activated state matches the local morphogen concentration will be described in this section. Another possibility, combining a sequential with a periodic mechanism to activate genes will be developed later (chapter 14).

The set of alternative states of determination may be compared with the steps of a staircase; a particular state would correspond to a wooden barrel resting on a particular step. It will remain stable on each of the steps but no intermediate levels are possible. The interpretation of positional information would correspond to a positioning of the barrel on a particular step under the influence of an external signal. A possible mechanism, by analogy, is that the barrel can be lifted up by a flood. The level at which the barrel comes to rest after the flood has diminished is a measure of the highest level of the flood (Fig. 11.4). In a system based upon a morphogen gradient the height of the flood is position-dependent and therefore, successively higher final stable states are attained at defined spatial intervals.

Molecular mechanisms enabling the controlled activation of particular genes

The selection of one particular state out of several alternative states requires competing feedback loops (Fig. 11.3). Genes and gene activators are the most obvious candidates for the required feedback-loops but it should be kept in mind that feedback can take place at other levels, e.g. in the control of translation or of RNA-processing and transport into the cytoplasm. In terms of genes, each gene i ($i = 1, 2 \ldots n$) has to have an autocatalytic feedback on its own activity and must compete with the others to assure that only a particular gene of the set can be active in any particular cell. To make such a system useful for the interpretation of positional information we have to arrange that the particular gene which becomes activated depends on the external signal, the local morphogen concentration. The following properties are required:

(1) Each gene i must have the tendency to activate the following gene in

the sequence, $i + 1$. In other words, a particular gene activity is triggered by the activity of the gene preceding it in the sequence.

(2) Transition to the next gene is possible only under the influence of the morphogen.

(3) The process of stepping through a sequence of genes must stop when the gene which is activated corresponds to the local morphogen concentration. Each step must require a progressively higher morphogen concentration. At a particular state of determination, the local morphogen concentration becomes insufficient to induce a further step. This requires some sort of hierarchy, similar to the levels of the steps in the staircase analogy given above.

The general principle can be molecularly realized in different ways (Meinhardt, 1978b). The following extension of eq. 11.3 (which describes the mutually exclusive activation of genes) should serve as an example. It allows a controlled activation of a particular gene under the influence of a morphogen:

$$\frac{dg_i}{dt} = \frac{c_i g_i'^2}{r} - \alpha g_i \quad \text{with} \quad g_i' = g_i + \frac{\delta m}{r} g_{i-1} \tag{11.4a}$$

$$\frac{dr}{dt} = \sum_i c_i g_i'^2 - \beta r. \tag{11.4b}$$

The first term describes the autocatalysis. A gene controlling a particular structure becomes slightly activated if the gene $i - 1$, the gene controlling the anterior or proximal neighbouring structure, is active. The cross-activation from the gene $i - 1$ is enhanced by the morphogen m and inhibited by the repressor r. The factor c_i describes the efficiency of the feedback. Arbitrarily, we will assume a hierarchy $c_{i+1} > c_i$. It is a property of the reaction eq. 11.4 that in the steady state $(dg_i/dt = 0)$ the concentration of the gene activator depends only on the decay rates $(g_i = \beta/\alpha)$ while the repressor concentration depends on which gene is active $(r = c_i \beta/\alpha^2)$. With the hierarchy chosen $(c_{i+1} > c_i)$, the repressor concentration increases with each transition to a higher gene. Since the repressor undermines the activation of the following control gene after each successful step, a higher morphogen concentration is required for a further step. The stepping forward will come to rest at a particular control gene which is determined by the m-concentration. Figure 11.5 shows a simulation of eq. 11.4 for a linear array of cells. The positional information is a smoothly graded function of the position of the cells. Nevertheless, the cells respond in an unequivocal way. In groups of neighbouring cells, the same gene is active and a switch from one gene to the next occurs without a zone of transition. Since the repressor concentration depends on which gene is active (Fig. 11.5) it can be used as a stable indicator for the achieved state of

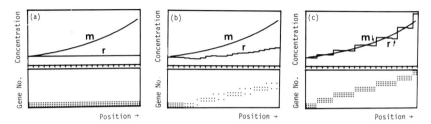

Fig. 11.5. Stages in interpretation of positional information. A set of feedback loops ("genes") is assumed which compete with each other via a common repressor (eq. 11.4). A sufficient morphogen concentration (*m*, positional information) enables the activation of the next control gene. This transition is connected with an increase in the repressor concentration (*r*, positional value). Since the positional information provokes the transition, the positional value slows down the transition, the stepping through comes to a rest if the achieved determination corresponds to the local morphogen concentration. (a) Initially, gene 1 is active in every cell. The activity of control genes is indicated by the density of dots. (b) Intermediate and (c) final stable state. In groups of cells, the same control gene is active and an abrupt switch from one gene to the next takes place between neighbouring cells. The pattern of gene activity would remain stable even after removal of the morphogen.

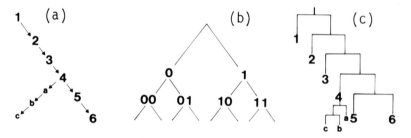

Fig. 11.6. Genealogical tree of determinations in systems controlled by positional information. (a) Under the influence of the morphogen, the cell switches from one state of determination to the next (1, 2, 3, 4, ...) controlling, for instance, which type of segment is to be formed in an insect. This suggests a ladder-like decision tree. The cells belonging to a particular specification can be further subdivided by a second positional information system, controlling, for instance, which part of an insect leg is formed (a, b, c, ...). (b) In contrast, a sequence of binary decisions, such as proposed by Kauffman *et al.* (1978). The sequence of decisions shown in (c) is formally equivalent to that shown in (a). Such a representation is used to characterize one type of cell lineages in Nematodes (Ehrenstein and Schierenberg, 1980; Kimble, 1981).

determination of the cell, as positional value. A control gene, once activated, remains active in a homeostatic manner even after a decrease of the morphogen concentration. The cell has a long-term memory with respect to the morphogen concentration to which it was once exposed. If, however, the morphogen concentration increases, the cells can switch to higher genes. In other words, if the positional information (morphogen concentration) is higher than the positional value (achieved state of determination, measured by the repressor concentration), the cells switch to higher states. If the positional information becomes lower than the positional value, a cell remains stable in its present state. This interpretation of positional information is a strictly local process. Cell communication is required only for the generation of the positional information, not for the response. If cells not normally neighbours are juxtaposed, missing structures of the gap are not intercalated as long as no new positional information is generated. Gaps which are not repaired can occur between insect segments (Fig. 8.3). The sequential switch from one state of determination to the next suggests a ladder-like sequence of decisions and not a series of binary decisions (Fig. 11.6).

Positional information in systems with marginal growth—the proximo-distal axis of the vertebrate limb

Pattern formation by reaction-diffusion mechanisms can occur with or without growth. In discussing systems in which cell determination is controlled in a sequential fashion by threshold effects of the morphogen concentration, we have assumed that no substantial growth occurs before the interpretation of positional information is completed. This assumption seems valid for the insect egg, and the antero-posterior organization of the vertebrate limb. In contrast, the proximo-distal axis of the vertebrate limb becomes determined during a period of substantial growth. The elements are layed down in a proximo-distal sequence (Saunders, 1948; Summerbell et al., 1973, for review see Hinchliffe and Johnson, 1980). If cells are determined by a local morphogen concentration, a source of a morphogen at the tip of the limb as such would be insufficient to account for the sequential formation of new structures, because newly formed cells at the tip would be exposed to the same morphogen concentration as previously formed cells. One may question whether a positional information scheme is realized at all in such outgrowing systems and what type of interaction would eventually allow the accretion of new structures during outgrowth. We have shown that a possible mechanism for the sequential formation of structures consists in the increase of the maximum morphogen concentration during outgrowth (Meinhardt and Gierer, 1980). Such increase can be accomplished by feedback of the achieved determination onto the morphogen production. In addition to sequential determination, the model provides an explanation for regeneration, for presence and absence of intercalary regeneration, and for the instability to form the most distal structures, the digits, without all proximal structures being present.

The bud of a vertebrate limb consists of mesodermal cells in an ectodermal jacket with a thickening at the tip, the so-called apical ectodermal ridge (AER, see also Fig. 10.7b). The AER is essential for limb outgrowth. Removal of the AER of the chicken wing bud leads to termination of further outgrowth. In this case only those structures which are already determined are formed. The later the AER is removed, the more distally complete the wing will be (Fig. 11.7). At a very early stage, limb bud mesoderm from the chicken can induce an AER even in an ectopic ectodermal cell layer (Kieny, 1960). Different models emphasize different aspects of limb development which should be part of an integrated explanation. The Saunders–Zwilling hypothesis (Saunders, 1969; Zwilling, 1961) stresses the mutual dependence of the limb bud mesoderm and the apical ectodermal ridge (AER). On the one hand, mesodermal cells are required to maintain the AER. The mesodermal cells presumably produce a substance required by the AER, the so-called apical ectodermal maintenance factor (AEMF). On the other hand, the AER induces the underlying mesodermal cells. To account for sequential specification along the proximo-distal axis, Summerbell *et al.* (1973) proposed that the cells in a so-called progress zone at the limb tip obtain, in the course of time, perhaps coupled to the cell divisions, a more and more distal determination. Cells leaving this labile zone are assumed to be fixed in their positional values. Faber (1976) proposed that a morphogen source is located at the tip but that the slope of the gradient close to the tip is so steep that it can be interpreted only after further outgrowth.

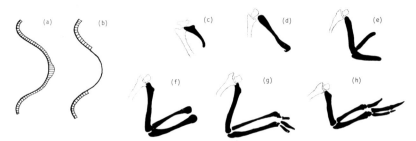

Fig. 11.7. Evidence for the sequential determination of the proximo-distal axis of a chicken wing (Saunders, 1948; Summerbell *et al.*, 1973; drawn after Summberbell, 1974b). (a, b) The operation: the apical ectodermal ridge (AER) is removed. (c–g) Resulting pattern after an AER removal at stages 19, 20, 21, 25. The later the AER is removed, the more complete the leg will be distally. (h) Bones of a normal wing.

We have provided arguments to show that the limb field results from the cooperation of patches of differently determined cells (Chapter 10). The AER marks presumably a boundary between "dorsal" and "ventral" cells (Fig. 10.3). In attempting to integrate the different aspects—the role of the AER as well as the origin and effect of a progress-zone—let us assume that the AER is the source of a morphogen which controls the proximo-distal axis. It generates a morphogen gradient with the high point at the distal tip. At a very early stage of limb development, only very few structures are determined under the influence of this incipient gradient, let us say, structures 1 and 2. To achieve the increase of the morphogen concentration during outgrowth, we will assume a feedback of the achieved state of determination onto the source strength. The mesodermal cells are assumed to produce a substance—we will also call it AEMF—which controls morphogen production in the AER. It is important that a more distal structure produces more AEMF. With the addition of more state-2 cells at the growing tip, the AEMF concentration at the AER increases and therefore the source strength of the AER increases as

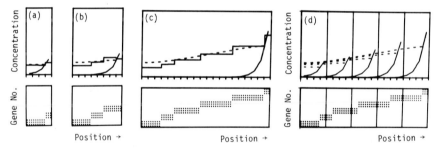

Fig. 11.8. Accretion of new structures in an outgrowing field. With outgrowth, the morphogen concentration increases gradually, due to feedback of achieved determination onto morphogen production, leading to a sequential activation of new control genes (lower half in each subpicture). The morphogen (positional information, ————) is produced only at the distalmost cell (the AER). The repressor concentration (steplike distribution) is a measure of the achieved determination and acts as the positional value. The diffusible AEMF (————), produced proportional to the repressor concentration, controls the morphogen production at the source, the AER. The initial morphogen distribution is sufficient to determine structure 1 and 2 (a). Due to growth at the distal side, the number of state-2 cells increases. This leads to an increase of the average repressor concentration and, via AEMF, to an increase in morphogen concentration at the tip. After sufficient growth, a switch into the next determination state is possible (b). At the distal tip only, positional information may exceed the positional value. More distal positional values can be acquired only there (c). Outside of this "progress-zone", in more proximal regions, positional information is always lower than the positional value and the cells remain stable in their once achieved state of determination. (d) A superposition of the distributions at different stages shows the increase in morphogen concentration at the outgrowing tip.

well. If a certain number of state-2 cells are present, the morphogen production of the AER becomes sufficient to switch some underlying mesodermal cells at the tip from state 2 to the state 3. Since AEMF is diffusible, its concentration depends on the average of the cell states, and the switch to state 4 is possibly only after a significant proliferation of state-3 cells has taken place. Only at a region close to the tip can positional information be higher than the achieved determination (Fig. 11.8). The model describes the progress-zone correctly. Cells at the tip acquire progressively a more distal determination while cells leaving this zone are fixed in their determination. Figure 11.8 shows a simulation of the pattern formation during outgrowth, in which the repressor concentration has been used as a measure of the achieved determination (see Fig. 11.5). The production of AEMF is assumed to be proportional to the repressor concentration while the production of the morphogen, taking place in the terminal cell only, is proportional to the AEMF. It is an inherent property of the assumed switching mechanism that determination of a cell can only be changed towards more distal determination and this only if the positional information is higher than the positional value.

Regeneration of structures of the proximo-distal axis of the vertebrate limb

The model describes not only sequential determination during outgrowth but also the regeneration of parts removed. After removal of the distal part of a limb bud, regeneration can take place after formation of a new AER (in amphibians) or after implantation of a new AER (in chickens). Since the mesodermal cells of the stump determine via AEMF the source strength of the new AER, the morphogen concentration will be similar to the corresponding stage of outgrowth. The missing parts can regenerate in the same way as the original pattern was formed. In agreement with experimental observation, the age of the transplanted AER is without influence (Rubin and Saunders, 1972).

If a very distal tip (giving rise to the wrist and phalanges) of an amphibian limb bud is grafted on a proximal stump, the intervening structures can regenerate (Fig. 11.9; Stocum, 1975a,b; Iten and Bryant, 1975). In more formal terms, if we denote the normal sequence of proximo-distal structures with 1, 2 ... 8, a gap of the type 123/67 will be repaired. In contrast, grafting a larger distal limb bud (3 ... 7) onto another stump at a distal level (1 ... 6) leads only to the structures expected from the fate map. In other words, the gap in the experimentally produced sequence 1 ... 6/3 ... 7 is *not* repaired (Fig. 11.9; compare with a "real" intercalation, Fig. 13.1). In both cases, the same structures (3 and 6) are juxtaposed. The difference in the intercalation indicates that the decision whether intercalation takes place or not, is not a local process. Remarkably enough, in the first case (123/67), the new structures (45) are derived entirely from the stump. The cells of the stump are therefore respecified towards a more posterior determination. According to the model, the distal cells of the transplanted tip still allow relatively high morphogen production. The gradient extends from the AER into the stump region. The cells of the stump become exposed to a higher morphogen concentration in comparison with their own determination. In agreement with the unidirectional interpretation of positional information, they switch to higher (more distal) determination. Due to this, more AEMF is produced, and this leads to increased morphogen production and finally to repair of the gap. The respecification is caused by a high morphogen concentration and

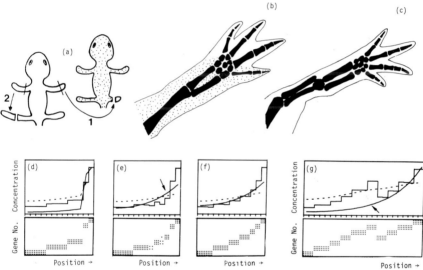

Fig. 11.9. Presence and absence of intercalary regeneration in amphibian limbs. If a distal fragment is transplanted onto a proximal limb stump (a, operation 1), intervening structures are reformed. The regenerate is entirely host-derived (b) as indicated by the pigment of the host (drawn after Pescitelli and Stocum, 1980). In contrast, if a larger distal fragment transplanted to a stump at a distal position (a, operation 2) no intercalation at the mismatching graft-host junction can be detected (c, drawn after Stocum, 1975b). Explanation in terms of the model: morphogen concentration etc. drawn as in Fig. 11.8. (d, e) After removal of intervening structures (d), the distal tip still allows relatively high morphogen production. Cells in the stump become exposed to higher morphogen concentration (arrow in e) and acquire a more distal determination. The gap can be repaired (f). In contrast, if the graft-host junction is more remote from the AER, (g) the morphogen concentration is lower (arrow) than the achieved determination and no respecification can take place.

not by a juxtaposition of normally non-adjacent cells. The absence of intercalation in a sequence 1 ... 6/3 ... 7 strongly supports this view. According to the model, after this operation, the graft-host junction is too remote from the AER. The morphogen concentration is insufficient for any respecification (Fig. 11.9c,g). The regeneration of intervening structures after implantation of a very distal limb bud fragment on a proximal stump and the absence of repair of a gap at a larger distance from an AER, provides the best evidence available that interpretation of positional information and not a mutual induction of neighbouring structures is involved in the determination of the proximo-distal sequence. In chicken wings, regeneration of intervening structures is possible only in very young wing buds, up to stage 22 (Summerbell, 1977). In terms of the model, whether internal deficiencies can be repaired depends on the range of the morphogen. If the range is small (low diffusion and/or short lifetime) the morphogen concentration declines rapidly with increasing distance from the AER (as shown in Fig. 11.8). After removal of an intervening structure the morphogen concentration in the stump area remains too small and no repair occurs.

A system of positional information with a feedback from achieved states has an inherent instability. More distal determination leads to an increase in positional information which, in turn, causes even more distal determination. The system is normally stable since AEMF transmitting the feedback diffuses and depends therefore on the average determination, while an increase in positional information leads only to a local distalization in the region close to the AER. An increase in positional information therefore has only a limited effect on the average determination. However, certain changes in the parameter can lead to an unstable situation. Examples are lowering of the AEMF diffusion rate or of the proliferation rate resulting, for instance, from killed cells. The result would be the formation of the most distal structures, the digits, at a premature position. Such a pattern has been observed in chicken limb buds after X-ray irradiation (Wolpert et al., 1979) and in children whose mothers have taken the drug Thalidomide during pregnancy (see Merker et al., 1980 for review).

A similar manifestation of the instability in the formation of the distalmost structure can be seen in the formation of digits if a small piece of a regeneration blastema is cultured in an ectopic position of the organism (Stocum, 1968). According to the model, cells at a distance from the AER must be available into which the increased AEMF can diffuse (Fig. 11.10), otherwise its accumulation would lead to premature distal transformations.

In conclusion, feedback of achieved determination onto morphogen production and therefore onto positional information provides a positional information scheme in outgrowing systems. The lability of cells in the progress zone and the stability at more proximal positions is a necessary consequence. With such a model, the mechanism for determination of the antero-posterior (Figs 10.1 and 10.7) and the proximo-distal axes become very similar and both axes become fixed by the cooperation of patches of differently determined cells.

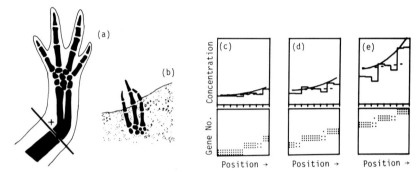

Fig. 11.10. Why do small fragments of a regeneration blastema preferentially form digits? After truncation at the level shown in (a), the first structures expected to be formed are those just distal from the cut (+). However, small pieces of blastemas cultured in an unusual position in the organism form the most distal structures, the digits (b, drawn after Stocum, 1968). (c–e) Model: for stability of the system some space is necessary into which the AEMF (– – –) can diffuse. Otherwise an increase in morphogen concentration (——) leads to a more distal determination which, in turn, leads to a higher AEMF and therefore to a higher morphogen concentration and so on until the most distal determination is reached. In agreement with the experimental observation, a somewhat greater cell mass would lead to a stable situation as can be seen from Fig. 11.9 which was calculated using the same constants.

12

Pattern formation by lateral activation of locally exclusive states

Several phenomena remain unexplained by the mechanisms of pattern formation discussed so far. These include the formation of a stripe-like pattern, the intercalary regeneration of missing elements in a sequence of structures, and the decision as to whether regeneration or duplication occur in imaginal discs. These phenomena are explicable under the assumption of a lateral activation of mutually exclusive states (Meinhardt and Gierer, 1980). An intuitive understanding of the mechanism envisaged may be provided by an analogy. Let us assume there are two families, A and B. At places where A is living, B cannot live and vice versa, they are locally exclusive. But both help each other and depend on the mutual help. A stable state is possible when areas populated by A are in close proximity to areas populated by B. The help has to be of a longer range, "across the street". Due to the required help, both can exist only in a close but well-balanced community. Due to the local exclusiveness, they belong either to A or to B but not to both and are therefore separated. The analogy is easily extended to more than two families, let us say $1, 2, 3 \ldots n$ and each needs the help of one or both neighbours. The most stable state is then a sequential order of the families in space.

The analogy can be used to illustrate the difference with respect to the mechanism of lateral inhibition discussed above which may be compared with the rise in power and wealth of one family, at the expense of the rest of the population. (For an adaptation of that model to socio-economic problems see Gierer, 1981c.) The population plays merely a passive role, it is the necessary background on which a centre of power develops in a self-enhancing manner. Such a family would engage all its power to suppress the

124

rise in power of a second family with similar ambitions, especially in a close community. In contrast, in the lateral activation mechanism, the co-existence of two (or more) different families is favoured since they need each other in a symbiotic manner. A mutual activation of cell types, a "cell sociology", has also been proposed by Chandebois, 1976b.

Molecular interactions enabling lateral activation

In the preceding section, molecular reactions have been discussed which lead to mutually exclusive states. The lateral help can be introduced via diffusible substances in a straightforward manner; several examples will be given. One is based on the reaction scheme drawn in Fig. 12.1; g_1 and g_2 are the (autocatalytic) substances required for the self-stabilization. The local mutual exclusion of the two states can be brought about by a common repressor (see eq. 11.3). The diffusible substances s_1 and s_2 provide the long-ranging help of one feedback system to the other:

$$\frac{\partial g_1}{\partial t} = \frac{cs_2 g_1^2}{r} - \alpha g_1 + D_g \frac{\partial^2 g_1}{\partial x^2} + \rho_0 \tag{12.1a}$$

$$\frac{\partial g_2}{\partial t} = \frac{cs_1 g_2^2}{r} - \alpha g_2 + D_g \frac{\partial^2 g_2}{\partial x^2} + \rho_0 \tag{12.1b}$$

$$\frac{\partial r}{\partial t} = cs_2 g_1^2 + cs_1 g_2^2 - \beta r \quad \left(+ D_r \frac{\partial^2 r}{\partial x^2} \right) \tag{12.1c}$$

$$\frac{\partial s_1}{\partial t} = \gamma(g_1 - s_1) + D_s \frac{\partial^2 s_1}{\partial x^2} + \rho_1 \tag{12.1d}$$

$$\frac{\partial s_2}{\partial t} = \gamma(g_2 - s_2) + D_s \frac{\partial^2 s_2}{\partial x^2} + \rho_1. \tag{12.1e}$$

Since all g molecules compete with each other, a disadvantage for one feedback loop is an advantage for the other. Therefore the lateral activation can be of a hidden form in which each of the feedback loops is subjected to a long-ranging self-inhibition:

$$\frac{\partial g_1}{\partial t} = \frac{cg_1^2}{rs_1} - \alpha g_1 + D_g \frac{\partial^2 g_1}{\partial x^2} + \rho_0 \tag{12.2a}$$

$$\frac{\partial g_2}{\partial t} = \frac{cg_2^2}{rs_2} - \alpha g_2 + D_g \frac{\partial^2 g_2}{\partial x^2} + \rho_0 \tag{12.2b}$$

$$\frac{\partial r}{\partial t} = \frac{cg_1^2}{s_1} + \frac{cg_2^2}{s_2} - \beta r \quad \left(+ D_r \frac{\partial^2 r}{\partial x^2} \right). \tag{12.2c}$$

(The equations for s_1 and s_2 are the same as eq. 12.1d,e.) The mutual help may be achieved by only one substance s. For instance, s can be produced by g_1 to which it is "poisonous" and can be destroyed by g_2 which needs it; g_1 needs g_2 for the removal of the poison whilst g_2 needs g_1 for the supply of s. Similarly, two stable states can be generated by two molecules repressing each other (eq. 11.2) and can be mutually stabilized by substances of a high diffusion range. A symmetrical form would be:

$$\frac{\partial g_1}{\partial t} = \frac{cs_2}{a + g_2^2} - \alpha g_1| + D_g \frac{\partial^2 g_1}{\partial x^2} \tag{12.3a}$$

$$\frac{\partial g_2}{\partial t} = \frac{cs_1}{a + g_1^2} - \alpha g_2 + D_g \frac{\partial^2 g_2}{\partial x^2}. \tag{12.3b}$$

(For the eq. of s_1 and s_2 see eq. 12.1d,e.)

In all these examples a homogeneous spatial distribution is unstable since, for example, a local g_1 elevation increases further due to the direct (eq. 12.1 and 12.2) or indirect (eq. 12.3) autocatalysis; the local g_1 increase is connected with a corresponding g_2 decrease (locally exclusive). Outside of this incipient g_1 maximum, g_2 wins the competition with g_1 due to the direct (eq. 12.1) or indirect (eq. 12.2) help via s_1. The result is an area of high g_1 (low g_2) and an area of high g_2 (low g_1) (Fig. 12.1).

These systems have features desirable for the explanation of properties of

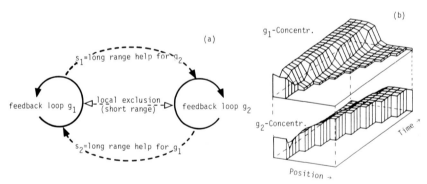

Fig. 12.1. Pattern formation by lateral activation of locally exclusive states. (a) General reaction scheme. Two (or more) autocatalytic feedback loops (g_1 and g_2) compete with each other, e.g. via a common repressor which leads to local exclusion. The long-ranging help ensures that both states are formed in a close and well balanced neighbourhood. (b) Simulation: g_1 and g_2 concentration is plotted as a function of space and time. A homogeneous distribution of both substances is unstable, an area of high g_1- and of high g_2-concentration is formed. The systems can show a good size regulation. In this example, proliferation of the g_2-cells leads to a corresponding enlargement of the g_1 area. Calculated with eq. 12.2.

different developmental systems. The interactions can subdivide a field into two or more parts with very good size regulation. For instance, if the g_2 area is relatively large in comparison with the g_1 area, g_1 is strongly cross-activated. Cells at the boundary are converted from the high g_2 into the high g_1 state until the correct proportion is restored (Fig. 12.1). The size regulation works only over a certain range, determined essentially by the range of the lateral help. If a field of cells has a much larger extension, a periodic alteration of g_1 and g_2 patches will be formed. Further, the mechanism allows the formation of a stable pattern in the short extension of the field, and of a stripe-like pattern. The observation of these features in isotropic developmental fields is a first indication that lateral activation may be involved.

Formation of stripes

Stripes—structures with a long extension in one dimension and a short extension in the other—are frequently encountered in development. A very obvious example is the colouration of many animals (see Murray, 1981), the stripes of a zebra being proverbial. In the visual cortex of vertebrates, areas connected with the right eye and with the left eye respectively are arranged in a stripe-like manner (Fig. 12.2). The thoracic segments of insects are subdivided into the stripe-shaped anterior and posterior compartments. Similarly, transplantation experiments with epidermal tissue of insect abdomen (Locke, 1959) suggest that the pattern elements have a very narrow extension in antero-posterior dimension but a large extension in dorso-ventral dimension. The pattern formation mechanism of lateral activation has the capability to form stripes. Since both feedback loops need each other in a close proximity, a long common boundary between both regions is favoured. Figure 12.2 shows a computer simulation of eq. 12.3. If the pattern formation is initiated by random fluctuations, the orientation of the stripes is somewhat irregular. Nevertheless, they consist of long narrow ridges. Perfect stripes are formed if some initial spatial cues are present, e.g. if the pattern formation starts at one side of the field. Both types of stripes can have a different width if the strength of the autocatalysis or of the mutual help is different in both feedback loops. Then, the equilibrium between the g_1 or g_2 cells would be shifted in favour of one or the other leading to a corresponding change in the number of high g_1 and high g_2 cells. In the ocular dominance columns mentioned above, visual deprivation of one eye leads to a narrowing of the corresponding stripes (Hubel et al., 1977). The question may arise why a stripe-like pattern emerges and not a checkerboard-like arrangement. In the latter case, even more boundary regions between "black" and "white" fields are created. Characteristic of a checkerboard pattern are sharp corners which

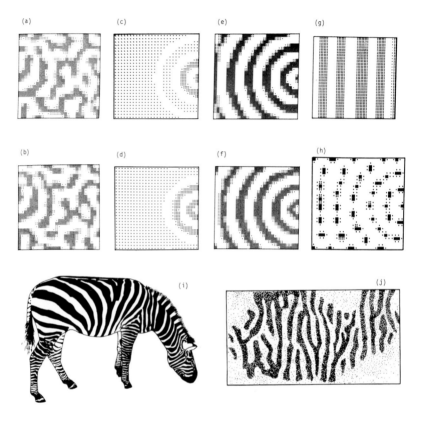

Fig. 12.2. Formation of stripes in a cell sheet. Mutual activation leads to areas of high g_1 or g_2 concentration (indicated by the densities of dots) which have a long extension in one dimension and a short extension in the other. (a, b) Complementary patterns after initiation by random fluctuation. The pattern is somewhat irregular but "mountain chains" can cross almost the whole area; Y-shaped branches frequently occur. Intermediate (c, d) and final (e, f) state in the pattern initiated at one point: perfect stable stripes are formed. Parallel stripes (g) result if a smooth gradient initiates the pattern. The lateral inhibition mechanism leads to patterns different from the mutual activation mechanism (h). The initially formed stripes decay into a bristle-like pattern. A computer program for these simulations is given on p. 195. (i–j) Examples of stripe-like pattern in biology. (i) Stripes on the coat of a zebra (see Bard, 1977). (j) Ocular dominance columns in the brain of a monkey (after Hubel *et al.*, 1977). The dark bands are regions innervated by the right eye, the light bands in between are connected with the left eye. In both patterns, the stripes show occasionally Y-shaped bifurcations similar as in the simulation (a, b).

(a)   (b)   (c)

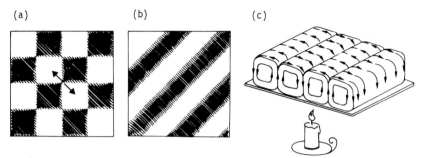

Fig. 12.3. Formation of stripes. Mutual activation favours long common boundaries between areas of high g_1 and g_2 concentration. A checkerboard-like arrangement is not favoured since it requires a high spatial resolution at the corners. To form larger coherent patches, some g_1 and g_2 diffusion is required, which blurs the resolution. A smoothing of the edges in the direction of the arrows (a) leads to the more stable stripes (b). Small asymmetries decide the orientation of the stripes. (c) Stripes in inorganic pattern formation: The Bénard-instability in a layer of liquid which is heated from below. The warmed up lower layer becomes lighter; long rolls of upstreams and downstreams are formed. Both exclude each other locally but enforce each other in a neighbourhood, satisfying our condition for stripe-formation.

would require a high spatial resolution for their formation. Any diffusion of g_1 and g_2 would blur this resolution. In contrast, a stripe-like pattern has no such corners (Fig. 12.3).

A non-biological example for such "stripe" formation is the arrangement of upstreams and downstreams in layers of liquids after the onset of the Bénard instability (see Velarde and Normand, 1980). Heated from below, the lower layer becomes lighter and tends to stream upwards, while the upper cooler layer tends to stream downwards (Fig. 12.3c). At a particular location either upstreams or downstreams are possible, but not both. They are locally exclusive. However, an upstream enforces a downstream in its surroundings and vice versa—it is obviously impossible to have only upstreams. Our formal conditions for stripe formation are therefore met in this example from physics, too.

In the following, the regulatory behaviour of some developmental systems is summarized and compared with that of the lateral activation mechanism, in particular with respect to size-regulation and the formation of striped pattern.

The dorso-ventral organization of the insect embryo

The dorso-ventral (DV) extension of an insect egg is only about one third of the antero-posterior extension. The possibility of stripe formation inherent in the mechanism of lateral activation, enables the formation of a stable high "dorsal" and a high "ventral" concentration along the whole antero-posterior axis and a graded concentration in-between. After a *longitudinal* ligation parallel to the A-P axis of a leaf hopper egg, a complete embryo is formed in both the dorsal and the ventral half of the egg (Fig. 12.4b). Each half produces many *more* structures when compared with the corresponding part of the non-operated egg. As shown in the simulations Fig. 12.4d–f, the mechanism of lateral activation is able to reform the terminal concentrations across the small (D-V) extension of the field even if, due to an experimental interference, it becomes even more narrow. By contrast, an activator-inhibitor mechanism would orient the pattern along the longest extension of the field. The employment of an activator-inhibitor mechanism for the DV-axis would require a primary organization of the A-P axis, e.g. a subdivision into segments as noted in Fig. 4.3.

On the basis of available experiments, it is difficult to assess whether the fine

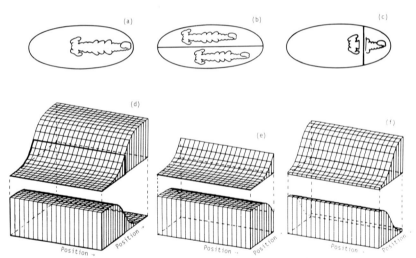

Fig. 12.4. Pattern regulation in the dorso-ventral dimension of an insect egg. (a) Normal embryo of a leaf hopper *Euscelis*. (b) After a longitudinal ligation, two complete embryos are formed, one in each fragment (Sander, 1971), indicating a high ability for regulation in the (shorter) dorso-ventral dimension. (c) No such regulation is possible in the antero-posterior dimension (see Figs 8.6 and 8.7). (d–e) The organization of the dorso-ventral dimension requires pattern formation along the shorter dimension of the field which can be accomplished by lateral activation of locally exclusive states. A narrow fragment (e), bordered in (d), leads to a regulation of the gradient in the even shorter dimension (*f*, see also Figs 13.7 and 13.8). That is in sharp contrast when compared with a pattern formed by a lateral inhibition mechanism where the gradient orient itself along the longest dimension of the field (see Fig. 4.3).

structure within the dorso-ventral dimension results from a concentration gradient, specifying positional information and leading, directly, to a position-dependent cell determination as discussed above for the A-P axis. Other mechanisms are conceivable. The DV-gradient can orient the sequence of dorso-ventral structures while this sequence itself is formed in a self-regulatory way (see Fig. 13.7). Or, the primary D-V pattern could determine only the terminal, most dorsal and most ventral structures whereas the other structures in-between are formed by intercalary "regeneration" (see Fig. 13.3). An indication in favour of a positional information scheme comes from a maternal effect mutation (*dl*) of *Drosophila*, isolated by Nüsslein-Volhard (1979). If heterozygoteous, the most ventral structure, the mesoderm is missing to a greater or lesser degree. The more dorsal structures are shifted and stretched towards the ventral midline. In a positional information scheme, missing structures are expected whenever the maximum concentration is not reached.

Compartmentalization and the re-establishment of compartment borders after experimental interference

The subdivision of the thoracic segments of insects into compartments (Garcia-Bellido *et al.*, 1973, 1976; Steiner, 1976; Wieschaus and Gehring, 1976; Crick and Lawrence, 1975) is presumably a key element in the understanding of the progressing subdivision of a developing embryo (see also Chapters 9 and 14). In recent years, much experimental effort has been concentrated on this subject and we would like to show that some basic regulatory features of compartmentalization can be explained by the lateral activation mechanism.

The thoracic segments have, at the time when they are determined, at the blastoderm stage, an antero-posterior extension of only 3–4 cells (Lohs-Schardin *et al.*, 1979). Almost simultaneously a clonal separation into anterior and posterior compartments takes place. These compartments have therefore the geometry of narrow stripes, 1–2 cells wide, which extend presumably in a belt-like manner around the blastoderm. As mentioned, the mechanism of lateral activation is able to | account for the stripe-like arrangement of two differently determined states. The connection of this compartmentalization and the segmental determination will become obvious in Chapter 14.

The compartments are characterized by the following features.

(1) *Clonal restriction.* A cell, once specified to participate in the formation of the anterior compartment will usually never be reprogrammed to form structures belonging to the posterior compartment.

(2) *Transgression of compartment borders.* After a severe experimental
interference with an imaginal disc, the compartment boundary can
appear at a new location. This shows that the normally fixed border
does not result from an irreversible determination but that it is
maintained by a dynamic process. Szabad *et al.* (1979) found after an
incision of the wing disc that the progeny of one genetically marked
cell participates in the formation of two compartments, indicating
unambiguously that some cells of the clone have been respecified (Fig.
12.5). Similarly, a fragment of the leg disc containing only cells of the
anterior compartment can regenerate structures belonging to the
posterior compartment (Schubiger and Schubiger, 1978). In these
experiments, the compartment border is only slightly shifted. Since the
structures formed are more or less normal, the shift can be visualized
only with genetic markers.

(3) *Compartmental respecifications.* After other types of experimental
interferences, the overall pattern is altered dramatically. For instance,
a heat shock leads in a mutant of *Drosophila* to some cell death and this
causes duplications or triplications of legs. Compartmental respecification (and with that, the formation of new compartment borders) is
presumably the primary event in this malformation (see Fig. 9.4). The

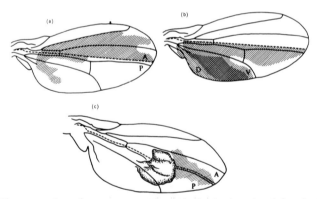

Fig. 12.5. Transgression of compartment boundaries in the wing (after Szabad *et al.*,
1979). Clones (hatched) are induced in a wing disc. The genetically marked cells are
the offspring of a single cell. One day later (day 6), either an incision is made in the
disc or cell death is induced. Due to these manipulations, the progeny of the marked
cells can populate different compartments. This would never occur without the
experimental interference. (a) A clone crossing the A-P boundary, (b) a clone
crossing the D-V boundary, populating the dorsal and the ventral wing surface. (c) A
wing with a bubble-like extrusion, crossing the A-P boundary. Such a distal structure
is expected if dorsal specifications appear in the ventral compartment (or vice versa)
close to the A-P boundary. It would be analogous to the leg duplication, Fig. 9.4.

location of the additional legs indicates that the cells of the outer anterior margin are especially labile with respect to a switch into a posterior specification.

These three phenomena—the normally fixed boundary, the possibility of a slight shift of the boundary, and a switch of some cells at a distance from the boundary into another compartmental specification—are easily explained but mutual activation of two locally exclusive feedback loops, A and P (g_1 and g_2 in eq. 12.1–12.3). The compartment border would be the transition between cells of high A and high P. The transition will be a sharp step if the substances accomplishing the self-stabilization, A and P, show very little or no diffusion. After removal of a large part of, for example, the high P area, the border is not shifted, because the help of the remaining P cells is sufficient to stabilize the A cells (Fig. 12.6a). However, after almost complete removal of the P area, the pattern formation process starts anew, leading to a new border at a different position (Fig. 12.6b). A higher diffusion rate of A and P leads to different behaviour (Fig. 12.6c). The compartment border is not as sharp and can be shifted if one compartment is too large in relation to the other, resulting in proportion regulation. Whether or not a sharp boundary between patches of differently determined cells exists may thus depend only on a difference in a diffusion rate and not in the underlying mechanism. The

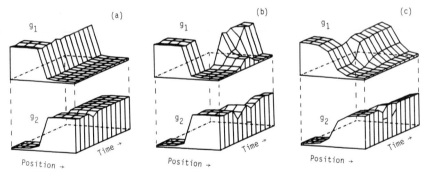

Fig. 12.6. Behaviour of a "compartment border" after an experimental interference. Assumed are two mutual exclusive states (g_1, top, and g_2, below), corresponding, for instance, to an anterior and posterior specification. The simulation shows a cross-section through a disc as function of time. (a) If the diffusion of g_1 and g_2 is low, the border between high g_1 and g_2 remains stable even after a substantial fraction of e.g. g_1-area is removed, the remaining g_1 cells stabilize the g_2 cells. (b) However, a complete removal of the g_1-area triggers the pattern formation again and a new compartment border is formed. (c) A somewhat larger diffusion of g_1 and g_2, the border is less sharp and a partial removal leads to a shift of the border, and to a new partition into two areas. Whether a boundary can be shifted or not could depend on small variation in the diffusion rate.

sharpness of the compartments in *Drosophila* and therefore their clonal restriction is presumably dictated by the small number of founder cells of a compartment. If many more founder cells were involved, a reasonable diffusion of A and P would be required to maintain these "compartments" as a contiguous patch of cells. Such diffusion would lead to a loss of clonal restriction. The possible absence of clonal restriction in other developmental systems does not indicate that different mechanisms are involved. Therefore, whether or not compartments are involved in development of vertebrates is presumably only a semantic question.

The possible reason for compartmental respecification after heat shock (and cell death) may be that the killed cells do not participate in the cell communication via diffusion. With that, the support of one cell type by the other may become insufficient and a switch to the alternative compartmental specification occurs. Figure 12.7 shows some altered pattern after induced cell death. The probability of respecification increases with distance from the border since these cells become less and less supported by cells of the other compartment. The details of this switch mechanism depend on how the lateral activation is molecularly realized (see eq. 12.1–12.3). For instance, if the bases were a mutual inhibition of competing feedback loops (eq. 12.2), any lowering of this inhibition can lead to a compartmental respecification. Such a process would be similar to the unspecific induction (Fig. 8.4). An increase in the basic production (ρ_0 in eq. 12.1 and 12.2), for

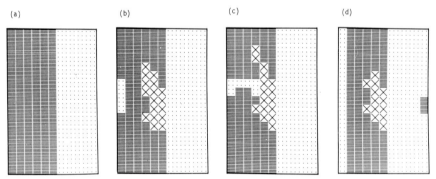

Fig. 12.7. The change of compartmental specification by cell death. (a) Two compartmental specification ($==$ and $:::$) are stable (see Fig. 12.6). (b–d) Killed cells (x) are assumed to participate no longer in the cell communication via diffusion. Hence, the mutual support of one state by the other may be reduced which can lead to a partial change in the compartmental specification. The switch from one determination to the other may occur in one compartment (b, c) or both (d), and at a position close or distant to the border, (c, b). A new compartmental specification can lead to a new system of positional information, causing a new proximo-distal axis (see Fig. 9.4).

instance, due to an elevated temperature, could also lead to a switch. On the other hand, the number of cells with A and P specification increases dramatically between the clonal separation (c. 20) and the mature disc (c. 50 000). This leads also to an increased distance of the cells from the border and, with that, from the other stabilizing cell type. The cells can be stabilized in the A or P determination if the stabilizing substances s_1 and s_2 are produced at a constant minimum rate (ρ_1 in eq. 12.1 and 12.2). A reason why A-cells are more easily reprogrammed to form P-cells than vice versa will be given on p. 163.

Systems with an organizing region at each end—regeneration of planarians

In discussing the regulatory features of hydra (Chapter 6), we have not considered that a hydra has in addition to the head a second organizing area with similar properties, the foot. Similarly, the head and the foot of planarians are two boundary regions which organize the field in between (see Chandebois, 1973, 1976a). In hydras as well as in planarians, a head, a foot or both regenerate even in very small tissue fragments, indicating that systems which bear an organizing centre at each end are stable over an enormous range of sizes. The head field and the foot field cannot be independent from each other, otherwise they would not appear at opposite sides. The mechanism of mutual activation of two feedback loops suggest an appropriate coupling of a head-forming and a foot-forming system which assures that both structures are present in the system and that they appear at maximum distance from each other.

In the application of lateral activation discussed so far, the common repressor which causes the local exclusivity has been assumed to be non-diffusible ($D_r = 0$ in eq. 12.1 and 12.2). This has the consequence that in each cell one and only one of the feedback loops is active (otherwise the repressor concentration would drop to such low values that one of the loops would become autocatalytic). This was appropriate to describe, for instance, the compartmentalization where a cell must be either anterior or posterior. In contrast, if the common repressor is diffusible, the autocatalysis of the two feedback loops g_1 and g_2 (the head and the foot activators) will be restricted to small patches of cells. In the rest of the cells, neither g_1 nor g_2 is produced. They are suppressed by the diffusible repressor. Since both loops produce the same repressor, they repel each other and the autocatalytic areas appear therefore at opposite ends of the field. Neither the head system can dominate over the foot system nor the other way round because of the mutual support of the two systems on a very long range (eq. 12.1). The same would be achieved if both systems, in addition to the common inhibitor, employ each a head- and a foot-specific inhibitor (eq. 12.2). Then, for instance, after removal of the foot, the foot inhibitor will drop until a new area of a high foot activator is induced. It appears at the opposite site of the head since this process is also sensitive to the common inhibitor. In Figs 12.8 and 12.9, these regulatory features are compared with those of planarian regeneration. The insensitivity with respect to size, and the ability to regenerate independent of whether one terminal structure remains present or not is in agreement with the experimental observation. Formation of a new head or foot does not require a complete separation of fragments; an incision may be sufficient for their induction (Fig. 12.9). The planarians also provide insight into how the antero-posterior and dorso-ventral axes are kept orthogonal to each other (which is not automatically the case if the patterns are formed by reaction-diffusion mechanisms). A condition for head or foot formation and thus for the re-establishment of an antero-posterior axis is a juxtaposition of dorsal and ventral tissue (Chandebois, 1979). A head or foot can therefore only be formed along the dorso-ventral borderline and never, for instance, within a purely dorsal region. This assures that both axes are orthogonal.

The same basic mechanism has been successfully applied to explain regulatory features of developmental systems which are so different as planarians and imaginal discs. In both cases, the common repressor (or inhibitor) keeps the two systems separate from each other and the long range

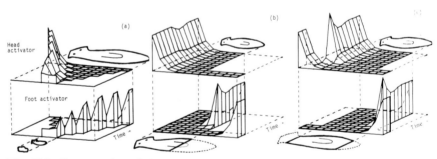

Fig. 12.8. Regeneration of planarians as example of pattern regulation in a bipolar field. Bipolar fields require two organizing regions, one at each end. It can be created by two activator maxima, one controlling, for instance, the head formation, the other the foot formation. A common inhibitor assures that both maximas appear at the largest possible distance from each other, at the opposite ends of the field. It is assumed further that on long range, both systems help each other either by a direct help (eq. 12.1) or by a specific self-inhibition (eq. 12.2). This assures that no system can suppress the other. (a) Pattern formation and growth: such pattern is stable over an enormous variation of sizes. (b) Removal of the foot leads to a regeneration of the foot activator despite the proximity of the head. (c) Simultaneous regeneration of the head and the foot activator. The result is independent of the precise position of the fragment.

help assures that both systems coexist with each other. Minor changes in the parameters—the substances bringing about local exclusion is diffusible or not—can account for the differences in these systems. In one case, the cells are either anterior or posterior and a sharp boundary exists in-between. In the other case, the head and the foot areas are restricted to the opposite ends of the field and both areas are separated by a region which is neither head nor foot. The similarities of both systems become apparent in the similar reaction upon the same experimental interference, an incision. In an imaginal disc this can lead to a compartmental respecification, in planarian to the formation of new heads and feet. This example demonstrates that by minor changes of a basic mechanism, an adaptation of its properties for different requirements of developmental systems can be achieved.

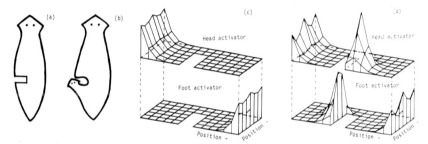

Fig. 12.9. (a, b) An incision into a planaria can be sufficient to initiate the formation of a new head and foot. (c, d) Model: in a two-dimensional field, a system with a head activator at one side and a foot activator at the other side would be stable. An incision, however, provides a diffusion barrier. The substances providing the mutual help (s_1 and s_2 in eq. 12.1 and 12.2) becomes so low that new maxima appear in the vicinity of the incision.

13

Generation of sequences of structures by mutual induction of locally exclusive states

A biological example: pattern regulation within a segment of an insect leg

In insects, the internal organization of a particular segment of a leg or of an insect abdomen has properties which differ essentially from the control of the overall sequence of the body or leg segments. To have a firm basis about what a theory has to explain, the main results of intercalary regeneration with cockroach legs (Bohn, 1965, 1970a,b; French, 1976a,b, 1978) should be summarized. For a brief description of the experimental results, the normal proximo-distal sequence of structures *within* a leg segment will be called 123456789. This assignment is somewhat arbitrary since clear demarcation lines such as the segment borders between different segments do not exist between the internal structures. It has turned out that:

(1) internal parts removed from a sequence of structures by surgical interference are replaced. For instance, an experimentally produced sequence 123/89 would intercalate the missing structures after one or two moults and the normal sequence 123*456*789 (intercalated structures are in italic) is restored (Fig. 13.1).

(2) Surplus parts are duplicated, e.g. an artificially produced sequence ...45678/456789 would form the structure 45678*765*456789 (Fig. 13.1c).

(3) A confrontation of the type ..678/345.. can also lead to an intercalary regeneration of the type ..678*912*345.. with an additional articulation in the intercalated 1-region (French, 1976a).

(4) The elements within the segments are specified in a repetitive manner.

Confrontation of different segments at the same internal level, e.g. a sequence **1234** of the femur (bold face) and a sequence 5789 from the tibia would not intercalate the missing elements **5678**91234 (see Fig. 9.6), while a confrontation **12**/89 does lead to an intercalation of the type **1234**56789.

(5) Intercalary regeneration is possible also in circumferential direction, removed longitudinal stripes are replaced (French, 1978, 1980). The pattern within abdominal segments show an analogous regulatory behaviour (Locke, 1959; Lawrence, 1966a; Wright and Lawrence, 1981a,b).

Possible mechanisms

Several mechanisms can be ruled out by just realizing that a sequence of the type ...56787654567... (Fig. 13.1) is stable. For instance, the sequence cannot be controlled by a concentration gradient, generated by a source at

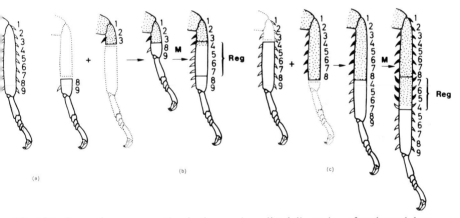

Fig. 13.1. Intercalary regeneration in the proximo-distal dimension of cockroach legs (Bohn, 1970a,b, 1971; French, 1976a). (a) The levels of the tibia are denoted (arbitrarily) with 1, 2 ... 9. (b) The confrontation of a proximal 123- and a distal 89-piece leads, after one or two moults (M) to the regeneration of the missing elements. By using different species or mutants (stippled, clear) it has been shown that most of the regenerate is derived from the distal elements, indicating a distal to proximal respecification. (c) Surplus structures become duplicated. The regenerate is again derived mostly from those cells at the mismatching junction which carry the distalmost determination. The spines of the regenerate have a reversed orientation, indicating that the sequence of elements determines the polarity of the individual cells and not the other way round. These experiments suggest a direct control of neighbourhoods and argue against long ranging positional information.

one end and a sink at the other end of the segment, since such a gradient would always be monotonic. The intermediate "maximum" would disappear. Is it possible to stabilize the intermediate "maximum", e.g. by an autocatalytic reaction as discussed above? Autocatalysis can compensate the loss by diffusion but would have the tendency to form the maximum concentration and thus to form the most terminal structures. We have seen such behaviour in the *Euscelis* egg where three abdomina (Fig. 8.2) can be formed. During intercalary regeneration, the elements forming the initial graft-host junction (8 and 4 in the example . . . 56787654567 . . .) are stable. An autocatalytic maintenance of the intermediate "maximum" is therefore unlikely. The same argument holds if one assumes that the gradient is stabilized by an active transport against the steepest slope (Lawrence, 1966a). The element 7 in a sequence . . 56765 . . . would profit from both sides and increase to 8 etc. until the terminal structure is formed.

An assumption which fits the observation more closely is that different qualities and not different quantities are characteristic for the particular elements of the sequence. The homeostatic property of the elements requires a self-stabilization. We will assume therefore that the sequence of structures consists of a sequence of differently determined (though perhaps closely related) structures, characterized for instance by the activation of particular genes out of a set of closely related control genes. The control of the correct neighbourhood of structures as indicated by the experiments summarized above, would require a mutual activation of the self-stabilizing states. Confrontation of cells which are usually not neighbours leads at the mismatching junction to a respecification of some cells into that of the missing structures (and presumably to an increased rate cell division, though proliferation is no logical requirement for the proposed mechanism).

The absence of intercalary regeneration when a proximal part of a femur is combined with the distal part of a tibia (**123456789** with **1234** from the femur, 56789 from the tibia) indicates that the same feedback loops are used repetitively for the determination of particular levels within different segments. Therefore only one set of few control genes would be required for the internal specification of different segments.

The ability to rebuild a removed part of an organism has a clear selective advantage. However, the removal of only an internal fraction from a leg or from an abdominal segment will never occur under natural circumstances. The presence of intercalary regeneration suggests that this process is not primarily invented to regenerate lost internal parts, but may be a normal process in the formation of the diversities of structures. For instance, during normal development, the terminal elements of the sequence could be determined and the sequence is then completed by the filling in of the missing structures.

Chains of induction

In the last chapter, we saw how two different structures can stabilize each other. The essential ingredients of the model are autocatalytic feedback loops which exclude each other locally but which help each other via diffusing substances. This mechanism can be extended to more than two structures (feedback loops) in a straightforward manner. The generalization of eq. 12.1 to many loops is

$$\frac{\partial g_i}{\partial t} = \frac{c_i g_i'^2}{r} - \alpha g_i + D_{g_i}\frac{\partial^2 g_i}{\partial x^2} \tag{13.1a}$$

with $$g_i' = g_i + \delta^- s_{i-1} + \delta^+ s_{i+1}$$

$$\frac{\partial s_i}{\partial t} = \gamma(g_i - s_i) + D_s\frac{\partial^2 s_i}{\partial x^2} \tag{13.1b}$$

$$\frac{\partial r}{\partial t} = \sum c_i g_i'^2 - \beta r. \tag{13.1c}$$

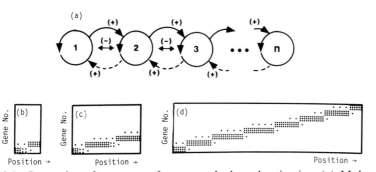

Fig. 13.2. Generation of sequences of structures by lateral activation. (a) Molecular interactions which allow the generation of self-stabilizing sequences of structures in space. Each state of the sequence 1, 2 ... has (i) feedback on its own, for instance via an autocatalytic gene activator, (ii) a long range activation of its neighbours and (iii) it produces and reacts upon a common repressor (◄——►). These three interactions lead to a self-stabilization and to correct neighbourhoods. (b–d) Simulation with eq. 13.1. Growth is assumed at the right margin. The concentration of the gene-activator molecules (shown as density of the dots) is plotted as a function of gene number and position. Initial separation of the field into two parts is accomplished as shown in Fig. 12.1. With increasing numbers of gene-2 cells, the concentration of the cross-activator of gene 3 (s_2, not shown) reaches a level which induces a transition of the gene-2 state into the gene-3 state etc. Long sequences of structures can be formed which are able to intercalate missing parts (see Fig. 13.3). (Equation 13.1 with $c_1 = 0.01$, $c_{i+1}/c_i = 0.74$, $\alpha = 0.1$, $D_g = 0.009$, $\beta = 0.15$, $\gamma = 0.1$, $D_s = 0.3$, $\delta^- = 0.4$, $\delta^+ = 0.12$.) (After Meinhardt and Gierer, 1980.)

In this example, the lateral help is introduced as a strong additional help and not as a necessary requirement (multiplicative factor). For instance, $\delta^- s_{i-1}$ describes the long range help from a lower neighbour. This enables stability of a particular state on its own, while in an interaction according to eq. 12.1 a state without a supporting neighbouring state would oscillate between the different states. The interaction according to eq. 13.1 has the capability for pattern formation, no external positional information is necessary to initiate the sequence. The formation of a sequence is shown in Fig. 13.2. The orientation of the emerging sequence depends on small asymmetries, e.g. any slight preference for the location of the second element in relation to the first is sufficient (see also Fig. 13.7). After the first two elements, No. 1 and No. 2, of the sequence have been laid down, the next state, No. 3, has to be triggered and not No. 1 again, otherwise only an alternation of two stages would emerge. A sequence will be formed if a state exerts a stronger help on the following state than on the preceding one. In terms of eq. 13.1, $\partial^- > \delta^+$. In fact, a term δ^- would be sufficient to activate each following state and therefore the whole sequence. As shown below, the term δ^+ facilitates intercalary regeneration.

Conditions for intercalary regeneration

Imagine a mismatching junction, for instance 12/678. Each structure has the tendency to induce its neighbours, especially the more distal neighbours. In the example, structure 2 tends to induce 3, structure 6 induces 7. Both structures cannot be formed at the border between 2 and 6 since the mechanism assures that the structures are locally exclusive. To achieve a correct regeneration of the sequence, the structure 3 has to be formed, i.e., the lower, more proximal structure must be dominating over a more distal structure. That signifies that a hierarchy exists among the feedback loops, or in terms of eq. 13.1 that $c_i > c_{i+1}$. For the communication between the cells a small diffusion of the g_i molecules is important. At the mismatching junction, g_2 and g_6 molecules are exchanged between the cells. Since the lower state 2 dominates, the g_6 production ceases in cells at the junction in favour of g_2. If structure 2 is sufficiently extended, structure 3 is induced (similar to during the original formation of the sequence) and the first step in the intercalary regeneration is completed. This process repeats itself until the correct neighbourhood of structures is restored. The mechanism is in agreement with the fact (Fig. 13.1) that both sides of the mismatching junctions contribute in the formation of the intercalate since the distal elements are reprogrammed by the contact with the more proximal structures and proximal structures form the more distal ones by the long-range induction. The mechanism has also the property of duplication of surplus

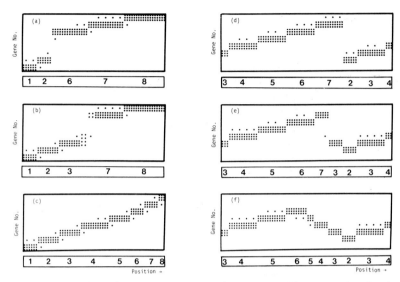

Fig. 13.3. Simulation of intercalary regeneration. Assumed is a mutual activation of locally exclusive states (Fig. 13.2). The repair of a gap is possible if the lower states dominate. (a–c) The area in which genes 3, 4 and 5 are active is removed (a). (b) Close to the mismatching junction, the activity of gene 6 will cease in favour of either state 2 or 3. (c) The missing structures are then reformed by cross-activation similar as in Fig. 13.2. (d–f) Intercalation of excessive parts in reversed polarity. The distally programmed cells become reprogrammed to form the proximal structures, in agreement with the experimental observation (Fig. 13.1c). The experimentally observed stimulation of cell proliferation at the mismatching junction is not taken into consideration. (After Meinhardt and Gierer, 1980.)

structures as shown in the simulation Fig. 13.3d–f since only the correct neighbourhood between adjacent cells is controlled and directional (vectorial) cell properties are not involved; the polarity of the sequence may be reversed during the correction of the neighbourhoods. The mechanism predicts that a particular element can only induce the neighbouring element. No averaging mechanism should occur. For instance a sequence 12/89 should regenerate via an intermediate state 123789 and not via 12589.

Organization of imaginal discs and insect legs around their circumference

A leg segment of an insect has a fine structure not only in its proximo-distal dimension, but also around its circumference (Fig. 13.4). From the very careful and detailed experiments of French (1978, 1980) we know that the

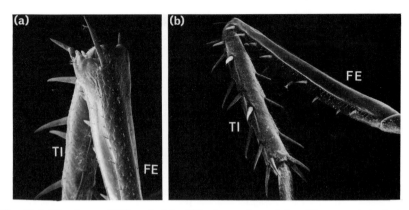

Fig. 13.4. The femur (FE) and tibia (TI) of a cockroach leg. The scanning electron micrograph shows the different structures of (a) the outer (anterior and posterior) and (b) the inner (ventral) face of the femur.

pattern around the circumference is able to intercalate missing structures or to duplicate excessive structures analogous to the pattern regulation in the proximo-distal axis. French *et al.* (1976) have proposed a "polar coordinate model" postulating that the circumferential pattern consists of a continuous sequence of structures to which they assigned arbitrarily the positional values 1, 2 ... 12. They are arranged like the numbers on a clock face. Missing structures are assumed to regenerate according to the rule of shortest route. Thus confrontation of the type 12/78 ... would lead to the regeneration of the missing structures 3456 while a confrontation ... 2345/234 ... would lead to the insertion of the structures 43, restoring in both cases normal neighbourhoods. This rule accounts for the regeneration-duplication phenomenon observed in imaginal discs (Bryant, 1975a,b). Small disc fragments duplicate the remaining structures while larger fragments regenerate the complete structure. For instance, a small disc fragment 2345 duplicates during the closing up and wound healing. The terminal structures 5 and 2 become juxtaposed and the structures 4 and 3 are intercalated. This leads to the (circular) duplication 2345432. A larger fragment consisting for instance, of the structures 23456789 would intercalate, according to the rule of the shortest route, the missing structures 10 11 12 1, leading to the regeneration of the complete sequence.

The polar coordinate model provides formal rules. What could be the molecular mechanisms on which this regulatory behaviour is based? How do the cells recognize what the shortest route is? In principle, the maintenance of the correct neighbourhoods of structures around the circumference can be achieved in the same way as described above for the proximo-distal sequence.

Long range activation of states which locally exclude each other leads to sequences of structures in which the correct neighbourhood is maintained in a self-regulatory manner. Since the circumference is assumed to consist of different qualities (not quantities as in a gradient system) each structure can support neighbouring structures: no special discontinuity occurs, for instance, between structure 12 and 1. Several questions are to be answered: how is such a sequence of structures initially formed in development? How is the circumferential pattern aligned with respect to the primary body axes? Why are left and right legs mirror images of each other? As we have seen (Chapter 9), the subdivision into compartments is the precondition that an imaginal disc and therewith a leg or any other appendage can be formed. Thus, in discs or in legs a coarse subdivision is given from the beginning. In the leg, these compartments are long narrow stripes running in proximo-distal direction. The compartments must be, so to speak, the frame for the finer subdivision around the circumference similar as the segment borders are the frame for the finer subdivision within the segments. In *Drosophila*, a particular tarsal bristle row coincides with the A-P compartment border (Lawrence *et al.*, 1979). The bristles are made irregularly from both compartments, indicating that along the border a signal is created (see Fig. 9.1b–d) which enables bristle formation. On the other hand, the symmetrical tarsal structures have no obvious relation to the non-symmetric pattern of compartments. Direct evidence exists that intercalary regeneration has something to do with boundaries. During intercalary regeneration of cockroach legs, boundaries of clonal restriction are maintained, suggesting a similar compartmentalization in cockroaches and *Drosophila* (French, 1980). For instance, cells of the posterior compartment can force cells of the anterior compartment to eventually regenerate missing structures up to the border, but the anterior cells will not give rise to posterior structures and vice versa. The question is then how the fine structured pattern of the circumference (e.g. 1–12) falls into register with coarse subdivision into the anterior (A), posterior (P) and ventral (V) compartment? A possible mechanism consists of a strong inducing influence of a particular compartment border on a particular structure, for instance the structure 1 may be induced by an A-P border, structure 5 by a P-V border and structure 8 by V-A border, followed by the intercalation of the missing structures (Fig. 13.5). This mechanism of coordinating the fine structure with the compartmental subdivision provides also a molecularly feasible basis for the regeneration-duplication phenomenon (Fig. 13.6). Small fragments contain, as a rule, only cells of two compartments. If, for instance, cells of anterior compartmental specification are missing in a fragment of a disc, they will remain missing. After closing the wound, a second confrontation of the remaining posterior and ventral compartment occurs and this leads to a duplication (Fig. 13.6a–c). In

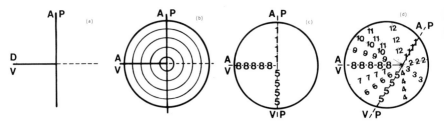

Fig. 13.5. Steps in the subdivision of an imaginal disc. (a) The primary event is assumed to be the formation of compartment boundaries. (b) By cooperation of three or four compartments a positional information system is formed (see Figs 9.1 and 9.2). The distance from the intersection of borders is decisive for the proximo-distal determination of the cells and whether the cells will form an imaginal disc or not. (c) Particular structures out of the set of circumferential structures are induced along the border between two compartments. (d) The missing structures are filled in by intercalation.

contrast, if some anterior cells remain in the fragment, a regeneration of the complete circumference would follow (Fig. 13.6d–f). However, the possibility of compartmental respecification has to be taken in consideration. A missing compartmental specification may be restored by a partial respecification of remaining cells (Figs 12.5 and 12.6). In such a case, both complementary fragments resulting from a partition of a disc can regenerate the complete set of circumferential structures such as observed by Kauffman and Ling (1981). In contrast, a duplication of both complementary fragments is less likely and would occur only after massive cell death.

French (1978) found a very striking absence of intercalation after confrontation of particular circumferential structures of the anterior with those of the posterior side of a cockroach leg. This situation is reminiscent of the absence of intercalation when, for instance, a mid-tibia is grafted onto a mid-femur. The missing distal femur and proximal tibia remain missing (Bohn, 1970a, see Fig. 9.7b). This has led to the conclusion that the positional values for the internal proximo-distal organization of segments are used in a repetitive manner within each segment and that the overall proximo-distal subdivision is made in a combinatorial way. A similar repetition of positional values within each compartment would explain why the confrontation of a mid-anterior compartment and a mid-posterior compartment eventually heals without intercalation.

In conclusion, like an umbrella needs at least three spokes to put up the interconnecting tissue, the three compartmental borders in the leg disc (or the two intersecting borders in the wing disc) are required to unfold the circumferential pattern. The final stabilization of neighbouring structures and intercalation of missing structures can be accomplished by long range

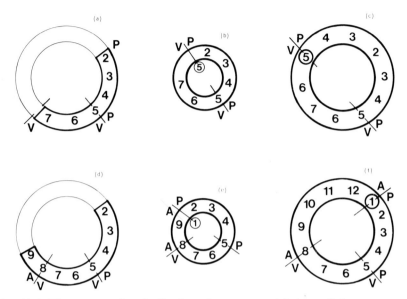

Fig. 13.6. The regeneration-duplication phenomenon. (a) A small fragment contains, as a rule, only cells of two compartments. In this example, the circumferential elements contain only structures belonging to the ventral and posterior compartment of a leg disc. In terms of Fig. 13.5, after closure of the wound, a new structure 5 (encircled) is induced at the newly formed P-V border (b). Intercalation of the remaining structures leads to duplication of the fragment (c). (d) If the fragment is only slightly larger and contains cells of the anterior compartment, an A-P juxtaposition results. This leads to the induction of a new structure 1 (e) and intercalation leads to the regeneration of all circumferential structures. The type of compartmental confrontation is assumed to be decisive as to whether regeneration or duplication occurs but the fact of transgression of compartment borders (see Figs 12.5 and 12.6) has to be taken into consideration.

activation and short range exclusion of different states. The control of the fine structure by the compartments ensures its correct orientation in relation to the main body axes. In connection with the model which describes the formation of compartments in the first place (Chapter 14) this mechanism accounts for the initial generation, the maintenance during further development and regeneration of the circumferential structures.

Sequence formation by induction and lateral inhibition

For the generation of a sequence of structures, we have assumed a long-range cross-activation of several competing feedback loops (eq. 13.1). An alternative would be that the size of each element is limited by a long-range self-inhibitory substance. Equation 13.2, a generalization of eq. 12.2, describes a possible interaction of substances:

$$\frac{\partial g_i}{\partial t} = \frac{c_i g_i'^2}{d_i r} - \alpha g_i + D_g \frac{\partial^2 g_i}{\partial x^2}$$

with
$$g_i' = g_i + m\delta^- g_{i-1} + \delta^+ g_{i+1} \tag{13.2a}$$

$$\frac{\partial d_i}{\partial t} = \gamma(g_i - d_i) + D_d \frac{\partial^2 d_i}{\partial x^2} \tag{13.2b}$$

$$\frac{\partial r}{\partial t} = \sum \frac{c_i g_i'^2}{d_i} - \beta r. \tag{13.2c}$$

A sequence of elements, generated in this way, has a good size regulation of the elements. If one element is relatively too large, the larger self-inhibition provides a disadvantage for that particular feedback loop compared with the other competing loops and it will shrink. For similar reasons, a very strong tendency exists to form each element of the sequence at least once in the field. Should an element be missing, the self-inhibition of the missing structure would become so low that it is induced via the cross-activation of the neighbouring structures (δ^+ and δ^- in eq. 13.2). Therefore, a sequence of the type 12/56 regenerates the missing elements 3 and 4. No hierarchy is required for this intercalation. However, the mechanism has no tendency to intercalate structures if this is connected with a duplication of existing structures as, for instance, in a sequence 2345/2345. The structures 4 and 3, missing at the gap, are already present twice in the field and the long-range self-inhibition emanating from the existing structures will suppress intercalation. The system has more the tendency to complete the two partial sequences. Without additional assumptions, this mechanism is not appropriate to explain intercalary regeneration within insect segments (13.1). However, as shown below, it may be the way to lay down the dorso-ventral structures of vertebrates.

Orientation of a self-regulating sequence by a gradient—an alternative to the interpretation of positional information

If a mechanism is given which has a strong tendency to form a sequence of structures in space (eq. 13.2) a small and possibly unspecific stimulus is sufficient to *orientate* the sequence. The sequence itself is formed in a self-regulatory manner. This offers an alternative to the measuring of local concentrations as discussed earlier for the interpretation of positional information (Fig. 11.5). Imagine a graded distribution of some substance or of a physical parameter which has, for instance, some influence on the cross-activation of the feedback loops (m in eq. 13.2). If initially the loop No. 1 is active in all cells of a field, the cells on one side switch faster to loop 2 and so on, the orientation of the emerging sequence is determined. Figure 13.7 shows that neither the steepness of the slope nor the absolute concentration but only the overall orientation of this gradient has an essential influence on the outcoming pattern. No special thresholds exist for the particular structures. Therefore, the orientating gradient need not be size-regulated for an adaptation of the correct size of the individual elements in relation to the total size of the field. The size regulation is a property of the mechanism which generates the sequence. In short, not the signal but the response would be size-regulated (Fig. 13.7).

If the orientating stimulus is symmetric, two sequences, mirror-symmetric to each other, can result. Each element is present twice in the field but each is

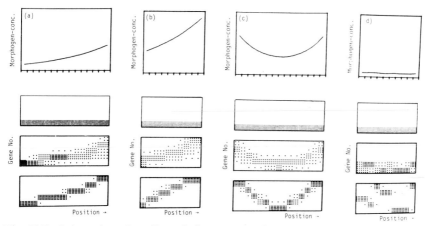

Fig. 13.7. Orientation of a self-regulating sequence by a gradient. If the mechanism for generating a sequence of structures (sequence of activated feedback loops or "genes") has the capability of pattern formation (reaction 13.2), a gradient can orient the sequence. (a, b) The emerging pattern is independent of the steepness or the absolute concentration of the gradient. The size regulation of the elements is a property of the sequence-generating mechanism, not of the gradient. (c) Initiation by a symmetrical distribution can lead to two complete sequences even if low concentrations are absent. Each element becomes correspondingly smaller. Such complete duplication is observed in the dorso-ventral pattern of amphibians (Fig. 13.8) and insects (Fig. 12.4). (d) If initiated by random fluctuation, the emerging sequence has an unpredictable orientation and gaps may occur but the mechanism eq. 13.2 assures that each element is present at least once in the field. From top to bottom in each subpicture: The orientating gradient (m in eq. 13.2), the initial, an intermediate and the final pattern of "gene activities" as function of position calculated with eq. 13.2 with $c = 0.01$, $\alpha = 0.03$, $D_g = 0.005$, $\beta = 0.05$, $\gamma = 0.02$, $D_a = 0.4$, $m\delta^- < 0.1$, $\delta^+ = 0.0$.

half as large. This type of pattern regulation is known to occur in amphibians. As shown by Spemann and Mangold (1924) in their classic experiment, the implantation of a dorsal lip of a blastopore into the ventral side of a blastula leads to a dorsal-ventral-dorsal duplication (Fig. 13.8). Cooke (1981a) has shown that the duplicated structures are squeezed into the same total field, the structures are correspondingly smaller. No additional cell proliferation takes place. Especially in small duplications, the structure next to the plane of symmetry—the pro-nephros—is frequently absent. According to the model, both pro-nephros would be relatively close together and one may suppress the other. By removing portions of the egg, Cooke has also shown that the complete dorso-ventral (D-V) pattern can be formed in a much smaller field which demonstrates the size-regulating features of the DV-pattern. For amphibians, it can be ruled out that the size regulation of the D-V structures result from a size-regulated DV gradient which is produced by a source-sink mechanism. No organizing properties of the ventral side can be detected upon transplantation.

A complete DV duplication is also possible in insects (see Fig. 12.4b). A comparison of these results with those obtained for the antero-posterior (A-P) pattern of insects (Figs 8.5 and 8.7) reveals basic differences between both systems. If a symmetrical (A-P) pattern is formed in insects, each half forms less structures in comparison with normal development. This has forced the conclusion that the local morphogen concentration controls which particular structure is formed (Fig. 8.5). In the D-V organization, each half forms many more structures—in fact the complete set—suggesting that in this case it is not the absolute concentration which is measured but that a self-regulatory sequence is triggered. It is remarkable that—as far as we know—in all systems in which the absolute concentration of a morphogen is measured (insect body

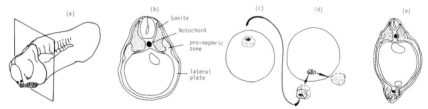

Fig. 13.8. Pattern duplication in the dorso-ventral axis of amphibian embryos (after Cooke, 1981a). (a) An embryo with the plane of the dorso-ventral cross-section shown in (b). (c) If a dorsal lip of a blastopore, the organizer, is transplanted to the ventral side of a blastula (d), a symmetrical duplication of the dorso-ventral pattern results (e). Both halves contain the complete set of structures. The individual structures are correspondingly smaller. That is very different from the antero-posterior duplication in insects (Fig. 8.5). The gradient produced by the organizer is assumed to orient a self-regulating sequence (Fig. 13.7).

segments, digits of vertebrates, segments of insect legs) the pattern to be formed consists of a repetition of similar but not identical subunits (see Chapter 14).

In conclusion, feedback loops which, on long range, support each other but on short range compete with each other can generate sequences of structures in space. These sequences are self-regulatory, missing structures can be added and gaps can be repaired by intercalary regeneration. The feedback loops could be, but need not be control genes.

Other applications of equations describing mutual activation of locally exclusive processes

For the explanation of the early evolution of genetic information, Eigen and Schuster (1978) have proposed equations similar to eq. 12.1 and 13.1. This similarity is not accidental. In the evolution of genetic information as well as in the activation of control genes, autocatalytic loops are assumed. In one case genes feed back on their own activity, in the other case pieces of nucleic acids are self-replicating. In both cases, a diversity of such competing loops should co-exist despite the "survival of the fittest". The co-existence results from the mutual dependence of the feedback loops. The essential difference in the model we proposed lies in the spatial order of the feedback loops which arises from their short range exclusion and the long range support. Further, each group of dividing cells in an organism represents an autocatalytic system. Different groups compete with each other since they consume the same nutritional substances. Nevertheless, the faster-dividing cells should not out-grow the others. A balance between the different cell types requires mutual dependence. A cancerous cell may have escaped this dependence from other cell types. Equations 12.1 and 13.1 describe essentially a type of symbiosis and the applications are presumably more general.

14

Digits, segments, somites: the superposition of periodic and sequential structures

A type of structure which is frequently encountered in higher organisms consists of a sequence of similar but not identical substructures (Fig. 14.1). For example, the segments of insects, separated by segment borders, are arranged in a repetitive manner. In the vertebrate limb, areas of presumptive digits and of programmed cell death alternate. However, each digit or each segment is different from the other, and is a member of a sequence of these substructures. Other examples are the bones of the backbone of vertebrates which originate from the sequence of somites.

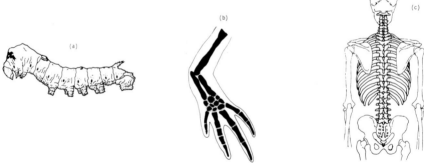

Fig. 14.1. Superposition of sequential and periodic structures—a basic pattern in higher organisms. Sequences of similar but not identical subunits form more complex structures. Biological examples: (a) the segments of an insect (a silkworm, drawn after Tazima, 1964), (b) the digits of an amphibian limb and (d) the vertebrae of a human being.

The assumption of a graded distribution of a morphogen and the interpretation of this positional information has enabled us to explain many experiments concerning the determination of the insect segments (Chapter 8), the segments of insect legs (see Figs 9.2 and 9.4) and of vertebrate limbs (Figs 10.1 and 10.7). In these models, the periodic aspect of these structures has been neglected. In this chapter, we will see how the periodic alternation between two or three alternative states enables in a gate-like manner the transition from one state in a sequence to the next in a very reliable way.

To see which type of mechanism can account for the generation of such dual structures we will again refer to the insect system, especially to *Drosophila*, since the most detailed experimental observations are available there. It will turn out that the mechanism derived from the insect system is able to explain observations in the somite system, indicating that the generation of sequential and periodic structures in precise register is a very basic mechanism in development.

The formation of the periodic pattern is the primary event

In the thoracic segments of insects, almost simultaneously with the clonal separation into segments at the blastoderm stage, a separation into anterior (A) and posterior (P) compartments takes place (Garcia-Bellido *et al.*, 1973, 1976; Steiner, 1976; Wieschaus and Gehring, 1976). Presumably they are arranged in a belt-like manner; each segment contains one pair of A-P-stripes. Both patterns are in precise register. For instance, the border between mesothorax and metathorax is also always a P-A border. Both patterns must arise in a coupled process. The question is then, which process is the primary event. Either initially the sequence of segmental specifications 1, 2, 3 ... is formed and each region is later subdivided into an A and a P region (1A, 1P, 2A ...) or the primary event is an A-P-A ... pattern and each pair of stripes obtains in a secondary process a particular segmental specification (Fig. 14.2). An answer to this very important question can be obtained from mutants in the control gene region responsible for the metathoracic specification, the Bithorax gene complex (Lewis, 1963, 1964, 1978; Sander, 1981). If, for instance, the locus *Cbx* or *bx* is mutated, a particular segmental specification extends into an adjacent segment without changing the A-P-A pattern (Fig. 14.3), indicating that the A-P-A pattern is independent of the segmental specification and that the formation of the A-P-A pattern is the primary event. It is rather the coupling of the segmental specification to this A-P-A pattern which is abolished in the mutation of the Bithorax gene complex (BX-C). In this chapter, we will see how this coupling is achieved during normal development and how particular transformations come about after a failure of particular elements in this switching system. That the

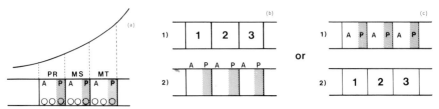

Fig. 14.2. Sequential and periodic pattern—what is the primary process? (a) In the thoracic segments of *Drosophila* the periodic pattern of anterior (A) and posterior (P) compartments is in register with the segmential specification (1, 2, 3 ...). The transition from the meso- to metathorax (MS-MT) coincides precisely with a P-A transition. This indicates that either the sequential pattern is the primary event and the A-P stripes are formed as a subpattern (b) or the formation of the periodic A-P structure is the primary event, and under its influence, each A-P pair gets a particular specification (c).

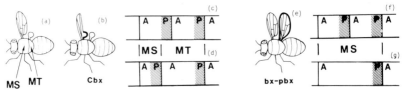

Fig. 14.3. Evidence that the periodic subdivision is the primary event. (a) Schematic drawing of a wild type of *Drosophila*. (b, c) In the *Cbx* mutant the region of metathoracic specificity (MT) is enlarged at expense of the mesothoracic (MS) region. Neither the A-P pattern nor the segmentation is affected (c). If the A-P subdivision were the secondary event, a subdivision of the smaller MS region into A and P and the subdivision of the larger MT region in only two compartments would be expected, as drawn in (d). (e, f) Similarly, in a *bx pbx* double mutation two pairs of AP stripes become MS specificity (f). If the A-P-pattern were a secondary subdivision, only one AP pair (perhaps larger) with MS specificity would be expected (g).

periodic subdivision is the primary event appears also reasonable from an evolutionary point of view. The insects evolved during evolution from lower Arthropodes and Annelides, creatures with many similar segments, indicating that the repetition of almost identical subunits was an early evolutionary achievement while diversification of the segments is a latter event.

"Gating" of the transition from one control gene to the next: the pendulum-escapement model

Assuming that the formation of the periodic structure is the primary event, a mechanism accounting for the precise superposition of both periodic and sequential structures must have the following features (Meinhardt, 1982): (1) it is able to create a stable periodic structure, possibly stripes; (2) the alternation of stripes controls the segmental specification; and (3) the number of repetitive elements and therefore the width of the stripes is under the control of a morphogen gradient.

The mechanism envisaged can be illustrated by an analogy. Imagine a grandfather clock. The weights are at a certain level (corresponding to the local morphogen concentration). They bring a pendulum into motion which alternates between two extreme positions. The escapement mechanism allows the hand of the clock to advance one unit after each change from one extreme to the other. The periodic movement of the pendulum is the primary event and the movement of the hand of the clock is under its control. As the clock runs down, the number of left–right alternations of the pendulum and hence the final position of the pointer is a measure for the original level of the weights (level of morphogen concentration). In terms of the mechanism for the interpretation of positional information, we will assume that, under the influence of the morphogen, the cell alternates between two states, to be called A (anterior) and P (posterior) and that the total number of alternations corresponds to the local morphogen gradient. Under the influence of this alternation, for instance with each P-A transition, the cell switches stepwise from one specification i to the next, $i + 1$ ($i = 0, 1, 2 \ldots n$). The stepwise advancement from one state to the other under the influence of the alternation between P and A may be compared with a ship in a channel with locks. A lock can be in two states. Either the upper gate is open and the lower gate is closed or other way round. In one state, the ship can enter into the lock but it can pass only after a switch into the other state. One state is characterized by the preparation, but blocking, of the transition. The other state enables the transition, but no entrance into the next preparative phase is possible. This enabling and blocking of transitions by the alternation between two states we will call "gating". Only a full cycle of alternation allows an advancement of one and only one step. In this way, the graded morphogen concentration becomes converted into the alternating A-P-sequence and into the sequence of structures $0, 1 \ldots n$. Since both patterns are formed in this coupled way, they are necessarily in register. A particular state of a cell can be characterized by its A-P state and its specification, for instance, 1A, 1P, 2A and so on. Cells exposed to a lower morphogen concentration obtain their final determination earlier, after a few alternations while distal determinations require more time. This is in agreement with the stepwise and unidirectional "promotion" of the cells under the influence of the morphogen which was concluded from the insect experiments (see Fig. 8.7).

The following elements are required for the realization of the model:

(1) The cells can be in one of two states (A or P). The transition from at least one of these states to the other, for instance P-A, requires a threshold morphogen concentration. The alternative transition (A-P) can be an autonomous process like the swinging back of a pendulum.

(2) The advancement from one specification, that means from one structure-controlling gene activity, to the next ($i \rightarrow i + 1$), proceeds under the influence of such a transition, e.g. P-A.

The oscillation between A and P and the generation of stable A-P stripes

The fact that an anterior fragment of a leg disc can regenerate posterior compartmental specifications indicates that the periodic arrangement of compartments is a dynamically stable system. We have seen (Chapter 12) how stripes can be formed and stabilized. The basic idea was that two states, to be called A and P, exclude each other locally but at long range both states help each other and depend on this help. This necessitates that both structures are formed in close proximity to one another. A stripe-like pattern is especially favoured since, in this case, the long common boundary regions enable an effective mutual stabilization.

While A and P cells stabilize each other in the region of a common boundary, it is a property of such an interaction that a group of cells consisting of one type only (A or P) can oscillate back and forth between the two possible states. If, for instance, all cells are in state A, the state P gets an enormous help while the state A is not supported. After a certain time, the cells switch from A to P. Later, the cells switch back to A for the same reason.

This spatially homogeneous oscillating system would be converted into a pattern which is stable in time if, at any location, an A-P border has been formed. Imagine a linear array of cells, which are under control of a graded morphogen concentration. All cells are in the P state and a certain morphogen concentration is required to induce the first P-A transition. Cells exposed at least to the threshold concentration switch from P to A and form in this way the first P-A border. Cells close to this border stabilize each other while the A cells distant to this border switch back to P, forming in this way a second A-P border. Again, cells distant to this new boundary will switch back to A, and so on. After each full cycle, one pair of A-P stripes is added. As the process progresses, the region of the stable spatially alternating A-P pattern enlarges at the expense of the spatially homogeneous cells which oscillate between A and P in time, The borderline between the stable and oscillating cells move over the field in a wave-like manner. This mechanism will continue until the total area is subdivided into a stable spatial periodic A-P pattern. A biological system in which repetitive structures are formed visibly in a wave-like manner is the genesis of somites of vertebrates. It will be discussed below in more detail.

Switching to new control gene under the influence of posterior-anterior (A-P) transition

After a stable state is reached in a particular cell, the number of oscillations a cell has made in its history is the same as to the number of the stripe it belongs in space (Fig. 14.4). The unequivocal correlation between the number of oscillations a cell has made and its position in the field enables a reliable activation of a particular control gene in a particular stripe, for instance, by the following mechanism. In the P state, the transition to next control gene is

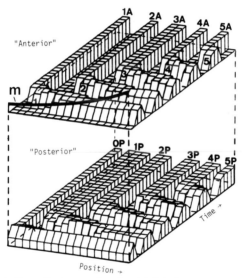

Fig. 14.4. The P-A-P oscillation under the driving force of a morphogen gradient. The simulation should demonstrate that the number of P-A transitions a cell has made in its developmental history is the same as the number of the A-P stripe in space in the finally stable pattern. Initially, all cells are assumed to be in the P-state. Those cells exposed to a certain morphogen concentration (*m*) switch to A; the first P-A border is formed. Those A cells distant to this border are not stabilized and switch back to P, forming a second (A-P) border. With each further oscillation, a new stable A-P stripe is formed. If for each further P-A transition a slightly increased morphogen concentration is required, the width of the A-P stripes is determined by the gradient (Fig. 14.10). The final result is a stable regular A-P-A pattern. Nevertheless, this A-P pattern has self-regulating features: An isolated patch of A-cells will re-establish an A-P pattern (Fig. 12.6).

prepared but the transition is blocked ("posterior block"). In the A-state, the transition is no longer blocked and the next following control gene becomes activated. However, no attempt is made in A to activate the next further control gene. As explained above with the ship and lock analogy, this has the consequence that a transition from one control gene to the next is possible only during a P-A transition and the control gene which finally becomes activated, depends on the total number of P-A transitions (Fig. 14.5). Let us assume that all cells are originally in the gene-0, P-state (0P). Only those cells which are exposed to a sufficient morphogen concentration will switch to A. Since it is a P-A transition, the cells switch from specification 0 (corresponding, for instance, to extraembryonal development in insects or to

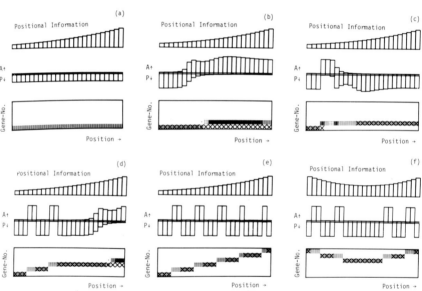

Fig. 14.5. Stages in the interpretation of positional information according to the pendulum-escapement model. The antero-posterior gradient (top of each subpicture) provides positional information. (a) Originally, all cells are in the P state and gene 0 (coding, e.g. for extraembryonal development) is active. In the P state, a substance (x) is produced which activates the next following control gene but this transition is blocked. (b) All cells exposed to a threshold concentration switch to the A state and therewith from gene 0 to gene 1 (since in the A-state the transition is not blocked). A and P cells need each other in the neighbourhood for mutual stabilization. Those A cells without P neighbours switch back to the P-state (c). For the next P-A switch, an increased morphogen concentration is required. This process will continue (d) until the total field is subdivided in the periodic pattern of anterior and posterior specifications and into the sequential pattern of gene activity. (f) Simulation of a "bicaudal" embryo (see Fig. 8.3). A computer program for these simulations is provided on p. 200.

the anterior necrotic zone in digit formation) to the state 1. The 0P and 1A cells stabilize each other, while cells further distant switch from 1A to 1P as described above. If the threshold remains unchanged, a periodic structure would be formed as described above since the next 1P-2A transition would take place in a region of even higher morphogen concentration. It is conceivable, however, that a certain incremental increase in the next P-A threshold results from the previous 0–1 transition. If this were so, a definite increment in the morphogen concentration would be required and the steepness of the gradient would determine the width of a pair of stripes.

Expected mutations and the phenotypes of the Bithorax complex of *Drosophila*

The best investigated complex of genes controlling a particular segmental specification is the Bithorax gene complex of *Drosophila* (Fig. 14.6; Lewis, 1963, 1964, 1978; Garcia-Bellido, 1977). At first glance, the phenotypes of the mutants appear quite puzzling. For instance, the anterior or the posterior haltere may become transformed into the corresponding part of the wing, flies with two wings can appear, the first abdominal segment may be transformed into mesothoracic structures and bear a fourth pair of legs, and so on. Why in most cases is only one half of a segment transformed? Why do these transformations respect compartment boundaries? Why do they cause, as a rule, a transformation into a structure of a neighbouring segment? The analysis of the phenotypes led me to the pendulum model as described above and after finding it, the mutations of the Bithorax complex (BX-C) appear to be the consequence of an underlying principle and not just an accumulation of genetic modification collected during the evolutionary history. It should be shown that the mutations are explicable under the assumption that the BX-C is the control gene for the metathorax (MT) and that its activation is gated by A-P-A changes. To see which type of mutants we expect on the basis of the model, the stepwise transition from one control gene to the next under the P-A alternation should be compared with the passage through a series of rooms. All the rooms are separated by doors. Each evening (P-state), one can proceed to the next door and ring the door bell. The next morning (A-state), the corresponding door will be opened. One can enter into the room and the door will be closed behind. However, one cannot proceed to the next door. This is possible only the following evening. The numbers of rooms someone has passed would correspond to the number of day-and-night cycles. The following types of "mutations" are expected:

(1) "Broken door": one can enter into the next room too early, already at the night before. If each room has two doors, one broken door is sufficient for an entry too early—the mutation is dominant.

(2) "Broken door bell": the door will not be opened correctly with the P-A transition. One remains in the last room. If two door bells are present, one is sufficient to ring the bell—the mutant is recessive.

(3) "Last door left open": one enters into the next room but the door cannot be closed behind. The new room has partially the character of the previous room.

The arrangement of the alleles on the chromosome as well as—according to the model—their normal function and transcription are shown in Fig. 14.6. Hayes *et al.* (1979) have proposed that the BX-C genes are transcribed from an operator region in the Ubx^+ region and that the direction of transcription depends on the compartmental specification of the particular cell. I will follow this proposal. Transcription is to the left (proximally) in the anterior compartment, thus enabling the transcription of Cbx^+ and bx^+

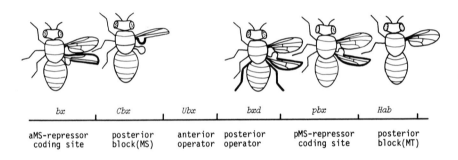

bx	Cbx	Ubx	bxd	pbx	Hab
aMS-repressor coding site	posterior block(MS)	anterior operator	posterior operator	pMS-repressor coding site	posterior block(MT)

Compartment and Segment	Genotype	Transcription bx Cbx Ubx bxd pbx Hab
P MS	wild type	PB ◄—
A MT	wild type	◄———————
P MT	wild type	PB ◄— ———————► PB
P MS	Cbx⁻	◄———————
P MT	bxd⁻	PB ◄— PB
P MT	Cbx⁻	◄——— ———————► PB
P MT	Hab⁻	PB ◄— ———————————

Fig. 14.6. The phenotypes of the mutants of the Bithorax gene complex (BX-C, Lewis, 1963, 1964, 1978) of *Drosophila*, their arrangement on the third chromosome and the normal function according to the model proposed. Abnormal structures are drawn with heavy lines. The arrows in the scheme below indicate the proposed transcription of the BX-C as function of the segmental and compartmental specifications. In the posterior state, the termination of transcription results from a posterior transcriptional block (PB). Mutation of the PB-sites (Cbx and Hab) can change the extent of transcription.

while in the posterior compartment, it is to the right, causing the transcription of bxd^+ and pbx^+.

To correlate the particular loci of the BX-C with particular functions in the model we have to compare the expected and observed mutations. In the PMS (posterior mesothorax) we expect that an attempt to activate the control gene for the MT, the BX-C, is made but that this activation is blocked, for instance by a transcriptional block of the BX-C. A failure of this block ("broken door") will be a dominant mutation and lead to MT structures in the PMS segment. That is the phenotype of the *Cbx* mutation (Fig. 14.7.). The Cbx^+ region is assumed to prevent the transcription of the *bx* region in a P-state. After a switch to the A-state, this posterior transcriptional block (PB) is released and *bx* can be transcribed. With this P-A transition, a transition from MS to MT specification should occur: we will assume therefore that wild type function of *bx* is to suppress the MS pathway. If *bx* is mutated (bx^-), a MS-repressor cannot be produced in the AMT segment. The AMT segment receives AMS character despite that the correct control gene is activated ("last door left open"). In contrast, in a Cbx^- mutant, the *bx* gene is already transcribed in the PMS segment, the MS repressor becomes produced and a PMS-PMT transformation occurs. With these assignments, the phenotypes of Cbx^- and bx^- are explained.

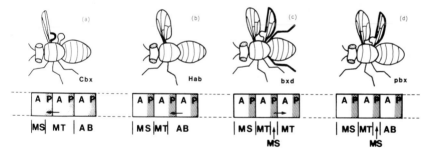

Fig. 14.7. Types of transformation expected from the model and observed in mutations of the Bithorax gene complex. (a, b) The posterior block does not work; the transition into the next posterior segment specification occurs already at the A-P transition. The segmental specificity is extended into the more anteriorly located segment. Corresponding phenotypes are *Cbx* (a) and *Hab* (b). (c) The activation of the next following control gene does not work correctly, the same segmental specification is repeated in the posteriorly located segment. This is the phenotype of *bxd*. (The formation of a fourth pair of legs is important in this context. The additional PMT-PMS transformation results from a particular arrangement of the loci on the chromosome.) (d) The correct control gene is activated in the correct region but the previously active gene is not suppressed. An example is *pbx*. Note that this is not an extension of a particular specification into a neighbouring segment.

After an A-P transition in the MT segment, the *Cbx-bx* transcription is blocked again at the *Cbx* region. A second coding region for the MS repressor, transcribed in the PMT segment, is required. This is assumed to be the *pbx* region. A *pbx* mutation therefore leads to a PMT-PMS transformation. The transcription of the *pbx* gene is assumed to start in the *bxd* region. A mutation in the *bxd* region leads therefore to a loss of function of a pbx^+ gene (Figs 14.6 and 14.7). In the PMT region, the activation of control genes responsible for abdominal (AB) structures must be prepared. In PMT, the transcription is assumed to proceed from *bxd* via *pbx* towards AB genes. However, in the PMT region the transcription of the AB genes is blocked by a second posterior block at the *Hab* region. In Hab^- flies, this block fails and abdominal genes are already activated in the PMT leading to a loss of the haltere and of the third pair of legs. Since Hab^- is of the "broken door type", it is dominant. In contrast, *bxd* is required to activate the AB genes; bxd^- is therefore of the recessive "broken door bell" type and leads to a repetition of thoracic structures in the first abdominal segment. In pbx^-, the correct control gene is activated but the product, the MS repressor, does not work; pbx^- is therefore of the type "last door left open". The gating mechanism is especially obvious in this part of the BX-C since the control genes are arranged on the chromosome in the same order as the corresponding structures in the real organism.

The model also describes the behaviour of double mutants. For instance, a fly carrying a *Cbx* and a *pbx* mutation shows a pure *Cbx* phenotype. It does not matter whether *pbx* is mutated or not. The defects are not additive. According to the model, due to the *Cbx* mutation, the MS repressor coding region at *bx* is also transcribed in the PMT region and the MS pathway is suppressed independent of *pbx*. This double mutation suggests that the bx^+ region is not a specific "selector gene" (Garcia-Bellido, 1975) for the AMT pathway but that it codes for a general MS repressor. The influence of the *bx* mutation is usually restricted to the AMT since, according to the model, *bx* is only transcribed there. The *Cbx* mutation is known to be an inverted insertion of the *pbx* region. I presume, however, that the mutant phenotype results not from this copy but from a destruction of a transcriptional block by this insertion.

The phenotype of double mutations frequently depends on whether the two mutations are located on the same chromosome (*cis*) or on different chromosomes (*trans*) and these differences are also correctly described by the model. For instance, if a *bx* and a *Cbx* or a *Ubx* and a *Cbx* mutation are located on the same chromosome (*cis*), the phenotypes are almost wild type although *Cbx* alone would be dominant. According to the model, if the transcription cannot start (Ubx^-) or the product is bad (bx^-) it does not matter whether the posterior block at *Cbx* works or not. In contrast, if bx^-

and Cbx^- or Ubx^- and Cbx^- are located on different chromosomes (*trans*), on one chromosome the transcription starts correctly (Ubx^+), it is not blocked in the PMS segment (Cbx^-) and the product (bx^+) is good. The normal Cbx^- phenotype results. Of, if bxd^- and pbx^- are in *cis*, the other chromosome can take over all functions and the phenotype is wild type. In *trans*, the transcription cannot start on one chromosome (bxd^-) while, on the other, the product is bad (pbx^-) and a pbx^- phenotype results, in agreement with the experimental findings (Lewis, 1963, 1964).

If, due to a chromosomal deletion, the BX-C is completely absent, the MT and all abdominal segments are of MS character (Lewis, 1978). In terms of the model, if the chain of sequential activation of control genes is interrupted at one step, the following genes in the sequence can be no longer activated. This does not mean that the BX-C genes are active in the abdominal segments. The activation of the BX-C can be a transient but necessary step in the activation of genes controlling abdominal structures.

The model allows one to understand other experimental observations it was not intended to explain. A striking asymmetry occurs in the regeneration of compartmental specificities. An anterior leg compartment can regenerate the posterior compartment but the reverse regeneration occurs much less frequently if at all (Schubiger and Schubiger, 1978). A similar asymmetry has been reported by Kauffman and Ling (1981) for the wing. In terms of the model, an A-P transition will occur whenever the stabilizing influence of the P region on the A state becomes too low. In contrast, a P-A transition would require the driving force of the morphogen and is therefore less likely to occur in an isolated disc fragment.

The assignment of very specific functions to the BX-C is necessarily speculative. Modifications are expected from a more complete understanding about how the segmentation proper is controlled (see below). Corrections may become necessary with the determination of the DNA sequence of the gene complex. It is hoped, however, that the general principle, the sequential activation of control genes by the alternation between two (or three) states, holds and facilitates an understanding of the information obtained from the sequencing of the DNA.

Sequential addition of new units at a zone of marginal growth

In some insects, only a fraction of the segments is formed directly. Then, in a second step, pattern formation is completed by adding new segments at a zone of marginal growth. The number of elements formed during the first or second phase is very different in different species (Krause, 1939). For instance, in *Euscelis* (Sander, 1976), only very few abdominal segments are added by growth, and the pattern formation can be essentially described as under the control of a morphogen gradient (see Figs 8.2 and 8.6). In contrast in crickets (Fig. 14.8), most of the segments are formed by marginal growth. Thus, some insects are organized by morphogen gradients during an essential non-growing period while others during a period of substantial growth. Smooth transitions exist between the two modes. This suggests that minor changes in the assumed mechanism should allow a pattern formation according to one or the other regime. That is the case in the proposed gating mechanism. Let us assume a growing marginal A-area. Whenever some A cells become too remote to stabilizing P cells, they switch from A to P (and vice versa). The switching can be used in the same way as described above to gate a transition from one control gene to the next. In Fig. 14.8, a simulation of a growing system is provided together with a biological example. Growing systems do not depend upon a gradient system which provides positional information since the order of the segmental specifications is determined by the growth.

Another pattern which is formed during marginal growth is the proximo-distal pattern of a vertebrate limb (see Figs 11.7 and 11.8). Summerbell *et al.*

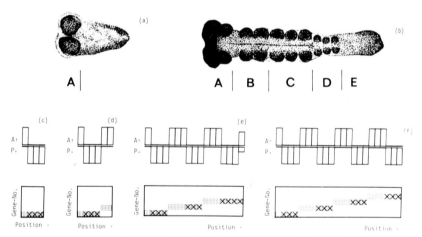

Fig. 14.8. Formation of a sequential and periodic structure by marginal outgrowth. (a, b) Biological example: stained germ band of a cricket. At an early stage (a), only the head lobe (A) is separated. Later (b), head segments producing mouth parts (B), the three thoracic segments with leg buds (C), and three abdominal segments (D) are formed. More abdominal segments will be formed in a sprouting-like process from the not yet segmented area E (after a photograph of P. Bader, see Sander, 1981). (c–f) Model: during outgrowth, whenever a particular state (A or P) surpasses a certain size, a switch into the alternative state (P or A) occur. Each P–A transition can cause a transition to a following control gene. If marginal growth is involved, no positional information is required. A computer program for this simulation is provided on pp. 200–205.

(1973) have proposed a progress-zone model according to which the dividing cells at the growing tip count the number of cell divisions and acquiring with each division a more posterior positional value. Cells leaving this zone of cell division maintain their once obtained positional value. On principle, such pattern formation can be also described by the gating model. The primary subdivision would not be the subdivision in the sequential structure (humerus, ulna etc.) but in a periodic structure (for instance, bone, joint, bone . . . or proximal, distal, proximal . . . part of a bone). Coupled to each (or each second) switch from one state to the other, a new specification in the sequential pattern is determined. However, the pattern regulation after removal of internal structures in the amphibian leg indicates that positional information is involved in the leg system (see Fig. 11.9b,e).

The formation of somites

A very important step in the antero-posterior organization of vertebrates is the formation of somites which give rise to the axial skeleton and musculature during further development. The paired somites are derived from two stripes of mesodermal tissue by a sequential separation into groups of cells. The separation progresses from anterior towards posterior. The periodic nature of somites is obvious (Fig. 14.9). Similar to the thoracic segments of insects, each somite seems to be subdivided into (at least) an anterior and a posterior portion since cells originating from the posterior half

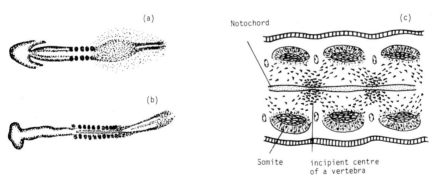

Fig. 14.9. Formation of somites and vertebrae. Somites are formed by clustering of cells. It starts at the anterior side, behind the head lobe and proceeds in posterior direction. (a) A chicken embryo at about 25h of incubation. Five somites are visible. (b) Ten hours later, 12 somites are present. (c) The anterior and the posterior part of each somite appear to be different. For the formation of vertebrae, cells from the anterior part migrate in anterior direction while cells from the posterior part migrate in posterior direction. Thus, cells from two different somites together form one vertebra (redrawn after Patten, 1958).

of one somite together with cells from the anterior half of the next somite form one vertebra (Fig. 14.9c). Presumably the somites are also different from each other since the vertebrae arising in this process are different from each other. Particular vertebrae form ribs while others do not.

Many experimental observations have provided insight into how the formation of somites is controlled. In amphibians, the first c. 20 somites are formed by a grouping of existent cells. Later, a graded transition to a more progress-zone-like addition of new somites in a region of cell proliferation in the tail bud occurs (Cooke, 1975a). The mode of segmentation is therefore similar to that in insects discussed above. Although the first somites appear long before the last somites, the size of the somites is regulated in such a way that the number of somites is almost constant and independent of the size of the embryo at the blastula stage. For amphibians, Cooke (1981b) has shown that only the first anterior somites (about 20) are size-regulated while the more posterior somites are always smaller but independent of the size of the embryo. The actual determination of the somites occurs earlier than their morphological appearance. Both processes can be experimentally distinguished by short heat shocks. In *Xenopus*, such a heat shock leads to defects of those somites which are formed at least 10h later. Meanwhile, five normal somites appear. This indicates that the (heat-sensitive) determination of somites precedes their appearance by approximately 10 hours. An important question is whether the determination or the morphological separation of a particular somite requires an inductive trigger from the previously formed anterior somite(s). Such a sequential trigger would explain the wave-like spreading of somite formation. This question has been answered by removing fragments from the posterior part of an amphibian embryo at the neurula stage. In this stage, the somites are neither determined nor visible. The surprising result is that in such fragments, the formation of somites takes place in the same sequence and at the same time as in the unoperated embryos. This indicates that the formation of one somite does not require its previously formed anterior neighbour. The actual somite formation occurs after a count-down-like process (Deuchar and Burgess, 1967; Pearson and Elsdale, 1979). However, in the heat shock experiments mentioned above, the number of malformed somites is much higher than expected from the shortness of the heat shock. Taking both observations together, the time at which the separation of the somites becomes determined and morphologically manifest seems to be cell-internally encoded. In the generation of fine structure, however, a neighbouring interaction seems to play an essential role in such a way that an irregular shape or a fusion of some somites has an influence of the successively formed somites. It requires the formation of several somites until the once evoked disturbance is smoothed out.

In summary, the model which should account for somitogenesis must have the following features: (1) A periodic structure is formed in an anterior to posterior order. (2) The individual somites formed in this process are different from each other. (3) Each somite is subdivided (at least) into an anterior and posterior part. (4) The size of the first anterior somites is controlled in relation to the total size of the embryo, the more posterior somites are of constant size. (5) The time at which the separation of somites occur is cell-internally determined. (6) Neighbouring interactions play a role in the generation of the fine structure.

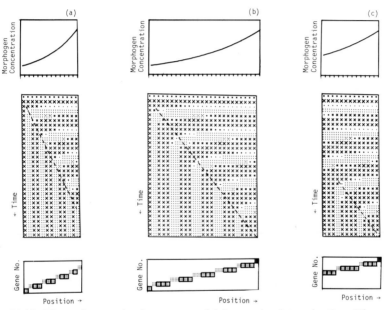

Fig. 14.10. The gating mechanism as a model for somite determination. The same mechanism as in Figs 14.4 and 14.5 is assumed. (a, b) A gradient with the same concentration range leads to a sequential and periodic pattern which is, within some limits, independent of the total size of the field. The upper subpictures show the assumed positional information, the central figure shows the oscillation between "A" (:::) and "P" (× ×) and the formation of a stable periodic A-P-A pattern. The broken lines mark the transition from the oscillating into the stable state and indicate, therefore, the moment of somite determination as function of time and position. The lower subpicture shows the final pattern of gene activities. A square indicates that the corresponding cell is in the P-state. The time required to form a somite does not depend on the size of a somite but on the A-P oscillation frequency. (c) A posterior fragment; no area of low positional information is present. The first A-P boundary can be formed only after a certain number of synchronous oscillation and switches to higher "genes". Separation into somites occurs therefore later, at the time corresponding to the local positional information (for computer simulations see pp. 200–203).

With minor modifications, the model for compartmentalization and specification in insects discussed above has all these properties. Cooke and Zeeman (1976) proposed a model for somite formation in which an oscillator gates a wavefront. In the model I propose, the oscillator (alternating between A and P), the wave front (separating stable and oscillating cells) as well as the spatial periodic pattern (of A and P) results from one and the same mechanism. In addition, the model I propose accounts for a different determination of the individual somites. As in insects, the oscillation seems to be under the control of a gradient with increasing concentration towards the posterior end of the organism. In vertebrates, this gradient appears to be relatively stable since the local values remain unchanged after isolation of fragments. The absolute level of this gradient would determine after which time or, more precisely, after how many A-P-A transition a stable boundary can be formed (Fig. 14.10). The formation of such stable boundary is assumed to correspond to the somite determination as discussed above. The steepness of this gradient would determine the size of the individual somites. (How a size-regulated gradient can be formed has been discussed above, Fig. 7.1.) If an increase of the threshold for the next P-A transition occurs only with the determination of the first (c. 20) somites, only these first somites will adapt to the steepness of the gradient. The activation of different control genes may be required only for the more anterior somites since they have to form structures with more individuality like ribs while the more posterior somites which forms the tail may be more or less identical. If the threshold is not increased, the A-P-A pattern is formed at the smallest possible distance. The simulations in Fig. 14.10 show the adaptation of the number of somites to different sizes of the field and that the somite formation can proceed normal in a posterior fragment.

The problem of segmentation

The question as to what is the signal to form a border between two segments or the cleft between two somites remains an open problem. Our assumption was that each of these repetitive substructures consists of an A and a P part. A segment border coincides with a P-A border. However, a juxtaposition of A and P cells cannot be the signal to form a segment border (or a cleft), since a second A-P confrontation is present in the centre of each segment, without a segment border being induced. Even if a pair of A-P stripes always has to form a segment, the grouping of a sequence APAPAP would be ambiguous; either AP/AP/AP ... or A/PA/PA/P ... would be possible. The internal polarity within the segments would not be determined.

The Bithorax gene complex has provided us with very important insights how segmentation is *not* controlled. As mentioned, several mutations (Fig. 14.3) cause the specification of a particular segment to extend into a neighbouring segment. The segment border no longer coincides with the transition from one segmental specification to the next. In other words, a transition from one segmental specification to the next, for instance from mesothoracic to metathoracic specification, can occur within a segment without a segment border being induced. We have to conclude that a segment border is not induced by a transition from one segmental specification to the next. After a complete deletion of the Bithorax complex, the metathoracic and all abdominal segments have mesothoracic specificity. However, the total number of segments remain unchanged. Thus, segmentation proceeds independently of whether the segments are different from each other or not. An assumption that more and more segments are added until the last abdominal segment is present is obviously incompatible with this observation. Counting of segments and giving them individual specifications are two different processes.

If neither the transition from one segmental specification to the next nor the P-A confrontation is the signal to form a segment boundary the question remains: what is the signal? One solution of this problem could be that the primary building blocks of a segment or a somite are not two states, A and P but three, for instance A, P and S (segment border). The primary periodic pattern formation would lead to an ... APSAPS ... sequence which allows a segmentation either of the type ... /APS/APS/ ... or ... /SAP/SAP/ ..., depending whether S/A or P/S induces a border. The advantage of having a subdivision into three parts is twofold. On one hand, the determination of the segment border is unequivocal, for instance SAP/SAP/ ... Secondly, the internal polarity of the segment is well-defined. No other grouping is possible. The sequences /SAP/ and /PAS/ have opposite polarities. We have seen in Chapters 12 and 13 how several structures in a sequence can be stabilized. The basic principle was that different states, for instance S, A and P, exclude each other locally but stabilize each other on long range. For instance, on long range S supports A, A supports P and P supports S and/or vice versa. This leads to a repetitive SAPSAP ... pattern.

Direct evidence for three such building blocks of a segment is not yet available but several experimental observations would find a straightforward explanation under this assumption. Nüsslein-Volhard and Wieschaus (1980) found a mutation in *Drosophila* in which twice as many segment borders are formed and in which the internal organization of the segments appears to be symmetric. Such a phenotype is expected if the central state is affected by the mutation. For instance, a /SAP/ pattern would lead, if the state A is affected, to a pattern /S/P/S/P/. Further, if an anterior and a posterior part of an abdominal segment of a bug is juxtaposed, a new segment boundary is formed (Wright and Lawrence, 1981a,b). The same happens after removal of a large internal part of a leg segment (French, 1976a). In the model, if the A area is removed from an /SAP/ sequence, the newly formed SP confrontation would induce a new boundary. The fact that a threefold subdivision has not yet been found is not an argument against this possibility. The A-P subdivision in the thoracic segments has been discovered due to the clonal restriction. However, the sharp AP boundary in the thoracic segments may be more the exception than the rule and required solely to define the coordinate system for appendages (Chapter 9). No clonal restriction is necessarily present between the other states. To the contrary, some diffusion facilitates the size regulation of the individual elements (see Fig. 12.6) and in the abdominal segments, no compartments have been found (Lawrence et al., 1978).

An open problem is further how the precise number of segments is controlled. The formation of the segments must be under the control of the primary gradient since the shallower gradient in double abdomen embryos (Figs 8.5 and 8.3) or in ligated eggs (Fig. 8.7) leads to fewer segments. A key observation are mutants in which each second segment is skipped, either the even numbered or the odd numbered (Nüsslein-Volhard and Wieschaus, 1980; Sander et al., 1980). A coherent model for this phenomenon is still missing.

Stepwise modification under the influence of A-P-A alternations

The activation of a next control gene under the influence of a P-A transition is only one of several possibilities. The Bithorax complex indicates that this possibility is realized in the insect system. Wo do not know how general this mechanism of activating particular control genes really is. Another possibility would be a systematic modification, for instance by a somatic processing of particular DNA or RNA sequences. The alternation between P-A-P ... could control the occurrence and the spatial distance of the modifications. The signal "modify" may require a simultaneous high A and P concentration. Only during the very short period of transition between A and P are both states active within the same cell since the states A and P mutually exclude each other. If an A-specific and a P-specific enzyme have to be present simultaneously to accomplish a particular biochemical step, this can take place only in the short phase of transition. A homeostatic maintenance of the once attained state would result since, after completion of the pattern, the cell would remain stably in one of the states and transitions would no longer occur.

The advantage of having a superposition of periodic and sequential structures

The model provides an explanation about how periodic and sequential structures can be formed. The gating and counting mechanism provides a mechanism to produce a large number of similar but different structures. Due to the superposition of the two patterns, the precision by which the morphogen concentration has to be measured is much reduced since the fine structure and correct neighbourhood emerges under the control of the periodic pattern with its higher spatial resolution. The compartments formed in this process are not only involved in gating the control genes. By cooperation of compartments, the A-P pattern can determine the position and orientation of appendages (Chapter 9). Alternatively, the A- and P-stripes may act as the terminal structures within abdominal segments and the missing structures are filled in by intercalation (Chapter 13). In any case, the superposition of both the periodic and sequential structures provides preconditions for making a reliable finer subdivision of a developing organism.

15
Formation of net-like structures

Net-like structures are common in almost every higher organism. The vascular system, the lymphatic system, the nervous system, the tracheae of insects, the veins of leaves and those of insect wings are examples (Fig. 15.1). Such net-like structures can be used to supply a tissue with nutritional substances, such as oxygen and water. The filamentous elements of a net consist of either linearly arranged, differentiated cells or of long fibres formed out of single cells. Filaments can provide information or mechanical stability and they can be used to remove certain substances from an area. A net-like

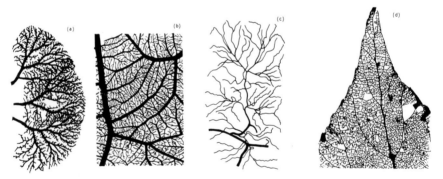

Fig. 15.1. Examples for net-like structures. (a) Dendritic trees of three nerve cells in a ganglion of the blowfly *Calliphora* (Hausen *et al.*, 1980, Figure courtesy K. Hausen). (b) Blood vessels in the allantois of the developing chicken. (c) Tracheal system in an abdominal segment of an insect (drawn after Wigglesworth, 1954, see also Fig. 15.3). (d) Skeleton of a poplar leaf. All other cells have been removed by microorganisms.

structure with all its ramifications is certainly not formed by the interpretation of positional information. This would require an enormous number of threshold values in each cell. Moreover, in many net systems, regulatory processes have been observed. For instance, new tracheae grow into a field of artificially evoked oxygen deficiency (Wigglesworth, 1954). Kühn (1948) found a mutant insect with a missing vein in the wing. The remaining veins were rearranged to compensate for the missing vein; there was no large gap in the pattern. Some substances are known to have an influence on the formation of a net: the plant hormone auxin in leaves (Jost, 1942), the nerve growth factor (Levi-Montalcini, 1964) on adrenergic nerves, a tumour angiogenesis factor (Folkmann et al., 1971; Folkman, 1976) on blood vessels. These findings indicate that elongation and branching are locally controlled processes. The question remains, as to what biochemical interactions govern the formation of elongated structures. How, for instance, can the elongation of a nerve be directed towards a particular target area? How is the very small surface area selected in which a new branch is initiated? Or, as in the case of leaves, how can the differentiation of cells into members of the vascular system proceed along a line, such that a certain distance from other vascular element is maintained? How are these processes encoded in the genes?

The model I have proposed for the generation of net-like structures (Meinhardt, 1976) is based on the repetition of two steps: (1) localization of the elongation of a filament, and (2) the elongation itself. The localization can be achieved by pattern formation based on short range activation coupled with long range inhibition, as described above. With regard to the orientation of elongation, let us assume that it is the purpose of a net to remove some substrate such as auxin or nerve growth factor from its surroundings. It could also be the remedy of some deficiency, such as the supply of oxygen mediated by insect tracheae. If autocatalysis depends on this substrate, the elongation will be oriented towards the increasing substrate concentration. The local high signal concentration causes an elongation of the filament which, in turn, causes a shift of the signal. Long filaments are formed as a trail behind a wandering filament-inducing signal. The mechanism will be explained in some detail for the special case in which filaments are formed by ordered differentiation within a field of undifferentiated cells. A generalization will be given later.

Formation of a filament

We wish to translate this idea into a mathematical model which can be interpreted on a molecular basis. We have to supplement the activator-inhibitor mechanism with a description of the differentiation process and a mechanism for the activator shift. We have seen (eq. 3.2) that a sharp local maximum can be generated by the interaction of an activator a and an inhibitor h:

$$\frac{\partial a}{\partial t} = c\,\frac{a^2}{h}\,s - \mu a + D_a\,\Delta a + \rho_0 y \tag{15.1a}$$

$$\frac{\partial h}{\partial t} = ca^2 s - vh + D_h\,\Delta h + \rho_1 y \tag{15.1b}$$

or
$$\frac{\partial h}{\partial t} = ca^2 - vh + D_h\,\Delta h. \tag{15.1b'}$$

(The new terms s and $\rho_0 y$ will be explained below. $D_a\,\Delta a$ denotes the generalized diffusion term for more than one dimension.) The local high activator concentration would be the signal for a cell to differentiate, to switch irreversibly from one state to another. The state of differentiation can be determined by the substance y; the concentration of y would be low in the undifferentiated state and high in the differentiated state. The transition from low to high concentration under the influence of the activator can occur in the following way. The activator produces y, but y also has a positive feedback on itself, which saturates at high y concentrations.

$$\frac{\partial y}{\partial t} = da - ey + y^2/(1 + fy^2). \tag{15.1c}$$

If, under the influence of the activator, a certain y concentration is attained, further increase of y is independent of the activator (see eq. 11.1; Fig. 11.1).

To get a filament, we have to arrange for the maximum to be shifted into a neighbouring cell. Let us assume that the purpose of the net is removal of a substrate s. Substance s is produced everywhere in the tissue at a rate c_0 and is removed by the differentiated cells at a rate εsy, while production of the activator depends on this substance s (eq. 15.1a).

$$\frac{\partial s}{\partial t} = c_0 - \gamma s - \varepsilon sy + D_s\,\Delta s \tag{15.1d}$$

Around each differentiated cell a depression in the s concentration will develop, but the s concentration increases steeply in neighbouring un-differentiated cells. In a newly differentiated cell, the s concentration decreases, slowing down activator autocatalysis. Activator diffusion into neighbouring cells can trigger a new activator maximum there due to the higher s-concentration. Due to mutual competition, only one of the neighbouring cells will develop a new maximum and even the previously active cell will be inhibited. The result is a shift of the activator maximum into a neighbouring cell which subsequently becomes differentiated itself. The next cell to be activated will be the one in front of the tip of this incipient filament. It is the adjacent cell with the highest s-concentration, because it has the least contact with the s-removing differentiated cells. By repetition of this process—shift of the signal, differentiation and shift again—long filaments of differentiated cells can be formed (Fig. 15.2). The structures which can be generated by this simple mechanism have features similar to those of biological networks. For example, bifurcations and lateral branches can be formed, the density of filaments can be regulated according to local demand, filaments can be oriented towards a target area and a damaged net can be repaired.

Formation of lateral branches

To form a net, individual filaments have to branch repetitively. A branch can be formed either by bifurcation at the growing tip or by the formation of a new growth point along an existing filament. According to the model, as the length of a filament increases, the inhibition, arising mainly from the

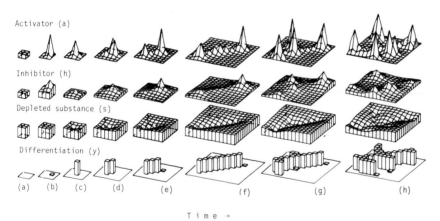

Time →

Fig. 15.2. Formation of a filament by differentiated cells and the initiation of a branch. By the interaction of an activator (a, upper row) and an inhibitor (h) a local high activator concentration is formed (a–b) which is used as a signal to differentiate the corresponding cell (c) (switching y from low to high concentration). The differentiated cell removes the substrate s which is produced everywhere. The high activator production, assumed to depend on s, escapes from the s-depression and is shifted to a neighbouring cell which is thereupon also differentiated. Indefinitely long filaments of differentiated cells can be formed by repetition of these steps. If the growing tip of the filament becomes sufficiently remote and enough space is available, the basic activator production of the differentiated cells can trigger a new activator maximum (f, g), which initiates a new branch. A computer program for such a simulation begins on p. 206.

activator concentration at the growing tip, may no longer be sufficient to suppress the basic (activator-independent, $\rho_0 y$ in eq. 15.1a, Fig. 15.2g) activator production of the cells of the filament. By autocatalysis, a new activator maximum may be formed along the filament, but, since the concentration of s is higher in the environment of the filament, the activator maximum is immediately shifted to a cell at the side of the filament and a branch is initiated.

Limitation of maximum net density

Net density will increase, since any branch can give rise to other branches. The ultimate net density can be controlled in two ways. In the first case, the activator, but not the inhibitor, is dependent on the concentration of s (eq. 15b'). Then, as net density increases, both the average s concentration and the maximum activator concentration decrease. Once a certain net density is reached, the activator maximum will be too low to induce further cell differentiation. The final net density will be proportional to the local production of s; net density will increase as long as more of the substance to be removed is present (see Fig. 15.8). This model may apply to the growth of tracheae into a region of experimentally induced oxygen deficiency (Wigglesworth, 1954). Alternatively, if production of both the activator and the inhibitor are dependent on s (eq. 15.1b), then elongation and branching will be independent of the absolute s concentration. Here, final net density can be controlled by a basic production of inhibitor by the differentiated cells ($\rho_1 y$ in eq. 15.1b). This creates a background inhibitor concentration in proportion to local net density. Filament elongation or the formation of new branches will cease if activator production is suppressed when the background inhibitor concentration rises above a certain level. This type of regulation leads to s-independent spacing of the net.

How a growing filament finds a particular target cell

According to the theory proposed here, elongation proceeds in the direction of the highest concentration of s. In the case of homogeneous s-production, that is usually located in front of the filament tip. If, on the contrary, the substance s is produced only in a particular area, the filament will follow the resulting gradient in s concentration upwards to the target area.

Regeneration of a net

Destruction of a filament of a net frequently leads to branching of nearby filaments, which repair the damage. Wigglesworth (1954), for instance, cut a

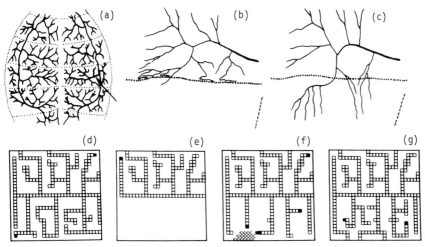

Fig. 15.3. Repair of a damaged net. (a) The tracheal system in the ventral abdomen of a normal, undamaged bug. (b) Oxygen supply to the fourth segment is disrupted by cutting the corresponding trachea (broken line, arrow in a). (c) Fourteen days later, tracheoles of the third segment have migrated into the oxygen-deprived segment (after Wigglesworth, 1954). In this process, patches of oxygen deprived epithelial cells send out cell processes which make connections with the tracheae. By retraction of the cell processes, the tracheae are pulled into an area of oxygen deficiency (Wigglesworth, 1959). Simulation: a complete net (d). After removal of the filaments in the lower half (e), new veins grow into the area (f). After complete regeneration (g), the newly formed part of the pattern look similar but not identical to the original net. The high activator concentration may be the signal for the epithelial cells to attract tracheae (differentiated cells: □, activated cells ■).

trachea which supplied a particular segment of a bug with oxygen. Tracheae from neighbouring segments subsequently migrated across the segment border (Fig. 15.3) and maintained the oxygen supply. The proposed model reproduces this regenerative capability. In an area without filaments, s is no longer removed. The increasing s concentration attracts new branches (Fig. 15.3) and the damage is repaired.

Formation of reconnections

In leaves most of the finer veins end blindly, but some of them connect with other veins to form closed loops (anastomosis). In the model, filament elongation is directed towards the largest available unfilamented space; a growing filament will therefore keep its distance from existing ones. A result of this mutual avoidance during growth can be seen in the final pattern of a leaf (Fig. 15.4a). In the model, this repulsion results from two different inhibitory factors on activator production. The inhibitor, centred around a growing tip and used to keep the activator localized, is a very strong factor. A weaker factor results from the depletion of the substance along an existing filament, which provides the stimulus for the shift of the activator peak away from the differentiated cells. A growing tip is strongly repelled by another growing tip, but it is only much more weakly repelled by an existing filament. This disparity in repulsive forces allows the avoidance mechanism occasionally to be overcome. Thus, when the weaker repulsion of an existing filament is overcome by the stronger mutual repulsion of two growing tips, reconnection of filaments is made possible. An example of such a reconnection is sketched in Fig. 15.4b.

According to Avery (1933), reconnections in the tobacco leaf are formed only after the transition from marginal to intercalary growth. That is understandable from the model since, during intercalary growth, two activator peaks can arise quite close together. As they develop more fully, strong repulsion will result from the increasing influence of their mutual inhibitors. They can thus be forced to elongate in directions that bring them into contact with other, older filaments. The midvein and the main lateral branches are formed before intercalary growth begins, which explains why it is that usually only branches of higher order form reconnections. For a complete description of plant venation, one has to take into account that the

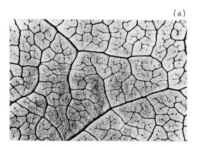
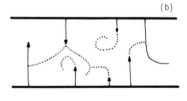

Fig. 15.4. Avoidance orientation and the formation of reconnections. The basic principle of the proposed model of line formation consists of repulsion between the differentiated cells and the differentiation-inducing signal. The elongation of a filament will be oriented away from other filaments into the largest available free space. (a) A record of this avoidance reaction can be seen in this maple leaf. However, reconnections (anastomosis) between veins are also possible. (b) According to the model, two growing tips (arrow heads) show a strong mutual repulsion, whereas the repulsion which an existing filament exerts onto a growing tip is more moderate. Connections are possible as the strong withdrawing movement (dashed lines) of two growing tips overrides the weaker repulsion arising from an existing filament. The number of reconnection depends therefore on the repulsion of an existing line and of how strong the tendency is to make a line straight.

veins not only remove the auxin from the surrounding tissue but that they transport it in a polar fashion towards the roots. In a young cross-vein, the polarity of the active transport is not completely fixed. Different parts of a vein can pump towards each other (Sachs, 1975). This would lead to local accumulations of auxin which, in turn, attracts other veins leading also to reconnections. The active transport of auxin itself seems to be also an autocatalytic process (Hertel and Flory, 1968). Mitchison (1980) has proposed a model for leaf venation which is based essentially on the transport of auxin.

The mode of reconnection described in Fig. 15.4 is only possible in a two-dimensional system. If a third dimension is available, the deflected filament would avoid the existing branches by passing underneath or above. Three-dimensional networks consisting of only one cell type, such as tracheae or lymph capillaries, usually end blindly. Extensive reconnections in three dimensions are possible, however, if two cell types are involved, as in the case of veins and arteries. Each cell type can form its own network by the repulsive interaction described above. Growth of the capillaries towards one other can occur if one cell type produces a substance which accelerates the activator production of the other cell type.

Variation in pattern formation

In the model, very complex patterns can be generated by the interactions of very few substances. These interactions can easily be encoded by genes. The question may arise as to how reproducible such patterns would be. The parameters determine only the general features of the pattern, such as average net density, the distance between branching points or the straight-ness of the lines. The fine details depend on external influences or even on random variations. For instance, during the initiation of a new branch, small differences between two neighbouring sites determine to which site the activator maximum will escape and therefore toward which site the new branch will grow. In a leaf, for instance, it does not matter whether the first branch leads to the left or to the right. Usually, such details are insignificant and can arise at random. However, once a branch has been made, let us say to the left, the resulting asymmetry strongly influences subsequent branching. Due to the presence of the new branch, the concentration of s will drop on the left site and the next branch will extend to the right, and so on. Since each decision depends to such an extent on the previous one, the overall pattern is reproducible. The random element in the process of pattern formation implies that two patterns generated by the same mechanism will be similar but not identical. This can be observed in nature. For example, leaves from the same tree are not identical, even though they are surely developed under

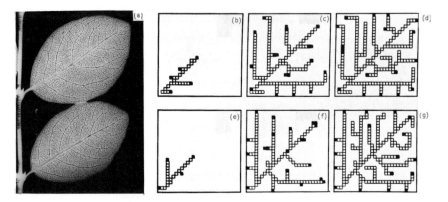

Fig. 15.5. Influence of random fluctuation on pattern formation. (a) Two leaves of the same tree. Their pattern formation is presumably controlled by the same genetic information. Nevertheless, the pattern is only similar, not identical. (b–d) and (e–g) Two simulations with the same parameters and the same initial condition but with different random fluctuation (3 %) in the constant c, eq. 15.1a,b. The model reaction determines only properties of the overall pattern such as average net density. Fine details are influenced by small local differences.

the control of the same genetic information. Similarly, the details of neuronal branching differ among genetically identical individuals of the water flea *Daphnia* (Macagno *et al.*, 1973). In Fig. 15.5, two simulations are provided, starting with the same initial condition but allowing different random fluctuations. The resulting patterns are similar but not identical.

Formation of a dichotomous branching pattern

One may ask whether this mechanism of line formation is the simplest possible. It is not. In this model, two inhibitory actions are involved in line formation, one to localize activation at the growing tip, and the other to determine the direction in which the centre of activation will migrate. As discussed in Chapter 5, the inhibitor may be replaced by a substrate which is depleted during activator production. Therefore, both tasks, the formation and the shift of the activator peak, can be mediated by one and the same substance s. Including the activator, only two substances would be sufficient to control differentiation (eq. 15.2):

$$\frac{\partial a}{\partial t} = ca^2s - \mu a + D_a \Delta a \qquad (15.2a)$$

$$\frac{\partial s}{\partial t} = c_0 - ca^2 s - \gamma s - \varepsilon sy + D_s \,\Delta s \qquad (15.2\text{b})$$

$$\frac{\partial y}{\partial t} = da - ey + y^2/(1 + fy^2). \qquad (15.2\text{c})$$

A pattern formed according to this interaction is given in Fig. 15.6. The main difference from the pattern formation discussed above is that here lateral branching in not possible. To provide sufficient drive, the differentiated cells have to remove such a substantial amount of s that the formation of secondary activator peaks along an existing filament is no

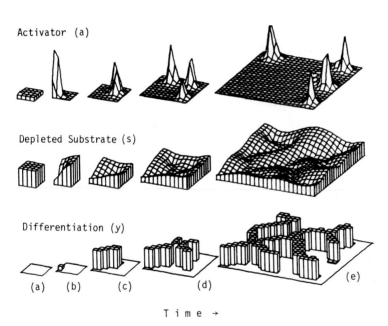

Activator (a)

Depleted Substrate (s)

Differentiation (y)

(a) (b) (c) (d) (e)

T i m e →

Fig. 15.6. Simulation of a dichotomously branching leaf pattern. The simple dichotomous leaf pattern shows only bifurcation of the growing veins, without later lateral branching. Such a pattern can be seen in the Ginkgo and some ferns (Fig. 15.7). Its simulation requires only two controlling substances. Local high activator (a, top row) concentration is formed by autocatalysis. Inhibitory action results from the depletion of the substrate s. (a, b) The local high activator concentration irreversibly differentiates (b) the corresponding cells (switching y from low to high concentrations). Since the differentiated cells also remove the substrate, the activator maximum wanders away from the differentiated cells (c). If enough free space is available, the activator maximum can split (see also Fig. 5.1), leading to a bifurcation (d–e).

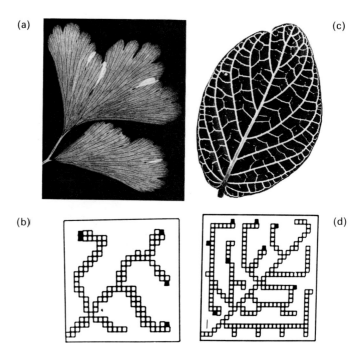

Fig. 15.7. Comparison of branching patterns of leaves with their simulations. (a) An evolutionarily older form of branching is dichotomous, such that a growing line can split into two, but lateral branching is not possible. (b) A simulation of dichotomous branching is possible under the assumption of only two controlling substances (see Fig. 15.6), in which the forking pattern is reproduced. (c) Lateral branching in a leaf of *Fittonia verschaffelti*; by a caprice of nature, the major veins appear white. (d) In the simulation initially one differentiated cell was assumed to be present (arrow). The first vein is oriented towards the largest available space, in the diagonal. The first lateral branches try to grow out at 90° but are repelled by the margins and establish, therefore, an angle of 45° with the midvein (see Fig. 15.5). The following lateral branches are repelled by the first and appear, therefore, also at 45° but branches of higher order grow out at 90°. The margins are avoided since the inhibitor cannot diffuse past the margin; it is, therefore, accumulated here. Reconnections are occasionally possible between higher order branches. Whereas the main lateral branches are quite straight, the higher order branches are more curved since they are frequently deflected by other growing tips. The simulation of a leaf with 29 × 29 cells can be only a crude approximation of the reality (even such a relatively simple simulation requires 28h on a relatively fast PTP 11/40 computer). None the less, it does demonstrate that such a complex pattern can be formed by the interaction of only a few substances.

longer possible. The activator maximum at the growing tip can still divide
into two maxima, allowing binary fission of an extending filament. It seems
that this is the evolutionarily older dichotomous| branching pattern of leaf
vascularization which can still be seen in the leaves of some ferns and the
Ginkgo tree. If this view is correct, the evolutionary step from the
dichotomous (Fig. 15.7a) to the common leaf pattern|(Fig. 15.7c) would
involve separation of the two inhibitory effects by the "invention" of a
separate inhibitory substance. The evolutionary advantage of lateral
branching is that it opens the possibility for intercalary growth. New
branches can be extended into the expanding spaces, providing necessary
nourishment to the tissue. In addition, damage to any particular vein is less
disruptive to the tissue as a whole, since the network has closed loops and
other pathways are available.

Filaments formed by oriented cell division or by extensions of single cells

Elongation of a line by accretion of newly formed, differentiated cells is only
one of several possible modes of line formation that can be simulated under
the proposed theory. The high activator concentration at the filament tip
could control cell division, while activator increase in front of the tip could
orient the process. In this case, a network is formed by organized
proliferation of the constituent cells.

In both the nervous system and the tracheal system, the elements of the
filaments consist of highly extended single cells. The formation of such a net
can be explained by the model under the assumption that a local high
activator concentration is formed on the cell surface. This can act as a
stimulus for the expansion of pseudopodes and the formation of a growth
cone. As in the orientation of a chemotactic sensitive cell (see Fig. 5.6), the
precise localization of the activator peak during the fibre elongation is
facilitated by periodic formation and decay of the signal. Harrison (1910) has
shown that the growth of a nerve fibre is not a continuous process but, rather,
that phases of fast elongation alternate with phases of searching for a new
direction. Several pseudopodes may be sent out at the same time, although
most of them will later be retracted.

A closer look at the branching pattern of nerves reveals features similar to
those seen in leaves. When the cells branch, the branches are typically at 90°,
and individual branches maintain distance from one another. In the growth
of a nerve, there again appear to be two types of inhibition involved. The first
localizes the growth cone and suppresses the development of additional
growth cones in surrounding elements. It also allows the selection of a very
small area where a new branch is actually initiated. The second inhibition

results from depletion of the substance s, which is possibly a nerve growth factor. This leads to orientation of each branch away from its neighbours; it is a mechanism for the mutual repulsion of nerve fibres. Genetically identical or similar individuals show the same general pattern of branching, but there are large differences in the fine structure of the branching (Macagno *et al.*, 1973). Of course, nerve cells do more than merely branch toward their targets; they also demonstrate spatial ordering of their connections. Any of the pattern generating mechanisms discussed above can supply the required spatial cues. Such spatial patterning can be used to set up a graded surface property in a field of the nerve cells. Together with the processes of competition and branching here described, this produces models that can generate many of the patterns seen in the retinotectal systems of lower vertebrates (Fraser, 1980; Gierer, 1981b).

Known substances which influence the formation of nets

As already mentioned, there are several substances known which influence the formation of particular nets. Some have been identified and others have been only partially purified. All these substances are comparable to the proposed shift-substance s, whereas none of the activator-inhibitor type are known. It seems likely that this is because substances of the shift-type s must be constantly produced or present in all the cells into which the filaments should grow. In contrast, the proposed activator-inhibitor substances are produced very locally and possibly only during short time intervals. Further difficulties in observing these substances result from the strong mutual influence of their production rates (see p. 46).

In leaves, additional vascular elements are formed after application of auxin to an injury (Jost, 1942). Auxin is known to be actively transported from the leaves to the roots. The veins remove auxin from their environment. Similarly, nerve growth factor NGF (Levi-Montalcini, 1964), which is necessary for the outgrowth of adreneric nerves, is actively transported from filaments to the cell body (see Thoenen and Barde, 1980). The highest NGF-concentration that a particular nerve will encounter is presumably that at its end, since more surface elements are available to remove NGF along a fibre. As long as no other constraints are imposed, this would lead to linear elongation of the fibre. The valley of NGF centred along each fibre would cause fibres of the same type to keep a certain distance from each other. If higher concentrations of NGF were present in a particular area, this area would attract growing fibres. According to the model, NGF and auxin would be co-factors for the activation of fibre elongation, for instance in the formation of a growth cone. The activation itself has to have autocatalytic properties, with an element of lateral inhibition.

The rapid growth of a tumour is only possible if it is extremely well nourished by a plethora of newly formed vessels (Algire *et al.*, 1945). Obviously, a malignant tumour must be able to overcome the body's control of blood vessel formation. Thus, an understanding of the control of vessel density is of great importance. Folkman *et al.* (1971) have isolated, from different tumours, a tumour angiogenesis factor (TAF) which induces ingrowth of vessels into normally avascular areas, such as the cornea or the epidermis (Fig. 15.8a,b). According to the model, the density of the net can be controlled by the substance *s*. Examples are given in Fig. 15.8. Low, medium and high *s*-concentrations lead to corresponding densities of the filaments. These densities may correspond to the situation of an avascular tissue, of a

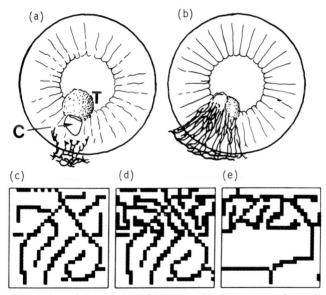

Fig. 15.8. Regulation of the density of a net. (a, b) Experiments by Folkman *et al.* (1971), redrawn after Folkman (1976). A piece of tumour tissue (T) and cartilage (C) are grafted into the cornea of a rabbit. After 20 days, some capillary growth is induced by the tumour but new vessels are inhibited in the vicinity of the cartilage. If the cartilage is inactivated by boiling, the tumour becomes vascularized (b), 30 days after the operation, the eye will be overgrown by the tumour. (c–e) According to the model, the density of a net can be regulated by the concentration of a substance *s* which is produced everywhere in the tissue and which is removed by the net. (c) "Normal" net density. (d) A two-fold higher *s* production in the upper half area leads to a much higher vessel density. This may be the situation in a tumour, with its excessive vascularization. (c) As (b), but an inhibitor-producing cell layer is assumed in the centre of the field; no vessels can grow through the inhibited area, but can circumvent it (eq. 15.1 with 15.1b′).

normal tissue and of a tumour. TAF may itself be the substance, or it may induce the synthesis of a substance, which is removed by the vessels and which is a co-factor in the local activation of elongation or the origination of a new branch. On the other hand, Folkman *et al.* (1971) have isolated a factor from avascular cartilage which suppresses the ingrowth of vessels. This factor could be the proposed inhibitor, since an externally supplied inhibitor can suppress activator production and so suppress elongation and branching, despite the presence of high s-concentration (Fig. 15.8).

16

Summary and conclusion: how to achieve the spatial organization of a developing embryo

Different models have been discussed for particular developmental systems and it may be worthwhile to summarize by showing how these mechanisms may be linked together to allow a reproducible development of an organism.

In the formation of the primary embryonic axis a process must be involved which is able to generate a pattern from more or less homogeneous initial conditions. A reaction in which a short ranging autocatalysis is coupled with a long ranging inhibition is able to generate such a pattern in a very reliable way. This mechanism accounts also for the re-establishment of an "organizing region" after an experimental interference or its unspecific induction. Small asymmetries in the environment of the maturing egg or of the early embryo can orientate the developing embryo in a predictable way. Influences as weak as gravitation (Kochav and Eyal-Giladi, 1971) are sufficient for such an orientation.

If the first pattern is involved to orientate, for instance, the antero-posterior axis, a second pattern would be required to organize the dimension perpendicular to the first, for instance, the dorso-ventral dimension. For the developing organism, it is absolutely essential that both patterns are perpendicular to each other (or at least not parallel). This can be accomplished by an appropriate coupling between the two systems, for instance if a border between "dorsal" and "ventral" is the condition to form the most anterior or posterior structure (or vice versa, see p. 136).

Patterns formed by reaction-diffusion mechanisms are necessarily transient since they depend on the size and the geometry of the fields. Both size and geometry change during development. The graded concentration profiles created by reaction-diffusion mechanisms can act as morphogen and provide

187

positional information for the cells. The cells change their (internal) state of determination in a systematic way until it corresponds to the external signal—the local morphogen concentration. This process may be in fact an oscillatory process. The cell may alternate between two states and this allows a "gate"-like transition from one structure-controlling gene to the next. It allows a counting of spatial structures on the gene level. Sequences of similar but not identical structures such as the somites or segments can be formed. Each element of such a sequence is necessarily subdivided into two parts. For instance an insect segment consists from the beginning of an anterior and a posterior compartment.

The determination of adjacent cells into different developmental pathways implies that borders are formed. Such borders enable further fine-structuring. Different cells on both sides of the border may co-operate through substances diffusing across the border, to produce together new morphogen molecules. The highest concentration is centred over the common boundary and the local concentration is a measure for the distance to that particular border. As we have seen, many experiments concerning the limb formation in vertebrates or of appendages in insects are explicable under this assumption. The dependence of the formation of new structures on borders between pre-existing structures assures a correct spatial relationship of newly-formed and existing structures. An arm cannot grow out of the hip region or the body cannot carry two left arms.

In principle, this mode of finer and finer subdivision of a developing embryo can be continued many times over. The interpretation of each sub-pattern creates new borders which, in turn, produce the pattern for the next subdivision. As far as we know for the insects, this mechanism is used twice; once for the primary subdivision into segments and secondly to form the segments of appendages. In the next further subdivision, the sequence of elements within a particular leg segment is formed. Such an intrasegmental pattern seems not to be controlled by a local morphogen concentration but is generated by a mutual activation of neighbouring structures. The different elements of the sequence stabilize each other on long range but exclude each other locally. Such a sequence of elements is dynamically stable over a substantial range of sizes and, after an injury, it can be repaired in a self-regulatory process.

Autocatalysis and lateral inhibition, early in development the driving force to initiate pattern formation, seems also to play an essential role during later development. It allows the spacing of repetitive structures such as bristles, hairs, feathers, stomatas or leaves. Depending on the mode of growth and initiation, the spacing can be more or less regular but in any case a minimum and maximum distance between the structures is maintained. This mechanism allows the selection of a small region out of a larger possible area. The precise location where a vein of a leaf should branch can be, for instance, selected in this way. Since this type of pattern formation is based on competition, small external influences can determine which region will dominate over the others.

The mechanisms of pattern formation and cell determination discussed above do not by themselves represent a complete theory of development. For instance, the very important question about how cell proliferation is controlled has been almost completely neglected in this book. Once an adequate theory of growth control is developed, interactions of growth, pattern formation and cell differentiation could be explicitly incorporated into the models.

It was our intention to show that the emergence of pattern during development can be explained by relatively simple, coupled biochemical interactions. All the ingredients used, such as diffusion and the mutual activation and inhibition of biochemical reactions, are known to exist in other biochemical systems. Explanations of a variety of phenomena have been given without additional unreasonable assumptions. Although unequivocal biochemical evidence of the existence of such pattern-forming mechanisms awaits future investigation, calculations have shown the internal consistency of the theory. Many models initially considered were found, by computer simulation, to be unable to account in a quantitative way for some initially chosen basic experimental observations. However, after a model consistent with these particular experiments was developed, it was also found often to be able to account for phenomena for which it was not originally designed. This of course, does not prove the validity of the model. None the less, it does suggest that the models are close to what actually occurs in development.

17

Computer programs for the simulation of pattern formation and interpretation

Computer simulations enable one to personally experience the pattern forming properties of the reactions proposed for the control of development. For this purpose, a selection of computer programs written in FORTRAN is provided together with some simulations. The general structure of the programs are similar: (1) input of the constants used for the simulation, (2) calculation of the initial conditions, (3) computing of the changes of concentrations and (4) printout of the new distributions. The computation consists of many steps (iterations) in which the concentration change within a short time interval is computed, the change is added to existing concentrations, and the resulting new concentration is used to calculate the next concentration change, and so on. The boundaries of the field of cells are assumed to be impermeable for the substances involved. This is incorporated in the simulations by setting a cell next to a boundary cell at the same concentration as the boundary cell itself. No flux takes place between two cells of identical concentration. Instead of printing long tables of numbers which are difficult to survey, the line printer is used to produce more graphic outputs.

Throughout the simulations, up to 20 constants are used. Their names and usual utilization are listed in Table 17.1. The first eight constants are integers and are used to select, for instance, the total number of iterations, the limits of the fields, or the type of equation. The remaining constants are used to control the diffusion and decay rates etc.

The input of these constants proceed via the subroutine CONST which allows an individual change of the constants during an interactive computer session (see Subroutine 17.5). To facilitate the understanding of the

programs, some of the printout sections are separated from the main
program and given in subroutines. One-dimensional and two-dimensional
arrays are plotted with the line printer by the subroutines PLOPP and XYPRIN
(Programs 17.6 and 17.7 respectively). A vertical line between outputs
indicate that they appear in an actual output below each other.

Table 17.1. Names and utilization of the constants used in the simulations.

No.	Name	Usual utilization	Corresponding Symbol		
1	IC	Total number of iteration	in equation:		
2	IPR	Printout after each IPR iterations	3.2	12.1	15.1
3	KX	Limits of the field	in program		
4	KY	Limits of the field	17.1	17.2	17.4
5	KZ				
6	IA	Selection of initial conditions			
7	IB	Selection of the reaction			
8	IC	Growth of the field?			
9	DIA	Diffusion of the activator	D_a	D_g	D_a
10	DIB	Diffusion of the inhibitor	D_h		D_h
11	TA	Decay rate of the activator	μ	α	μ
12	TB	Decay rate of the inhibitor	ν	β	ν
13	QA	Basic activator production	ρ_0	ρ_0	ρ_0
14	QAA	Basic inhibitor production	ρ_1		
15	QB				d
16	QBB				c_0
17	QC	Random fluctuation			
18	QD	Decay rate of the long range help		γ	ρ_1
19	QE	Diffusion of the long range help		D_s	γ
20	QF				ε

```
      C-----PROGRAM FOR SIMULATION OF PATTERN FORMATION
      C-----IN A LINEAR ARRAY OF CELLS
      C-----A(I) ACTIVATOR, B(I) INHIBITOR, Y(I) SOURCE DENSITY
0001        COMMON/D/ IC,IPR,KX,KY,KZ,IA,IB,IZ,DIA,DIB,TA,TB,QA,
            1QAA,QB,QBB,QC,QD,QE,QF
0002        DIMENSION A(31),B(31),AD(31),BD(31),Y(31)
0003        RRN=RAN(0,0)
0004    150 CALL CONST
      C-----INITIAL CONDITION
0005      1 AFA=TB/TA+QA/TA    !  CONZENRATIONS OF THE SEMISTABLE EQUILIBRIUM
0006        BFB=AFA**2*.01/TB
0007        WRITE(5,907) AFA,BFB
0008    907 FORMAT (/,' HOMOGENIOUS ACTIVATOR- AND INHIBITOR CON. ',2F8.5)
0009        DO 140 I=1,31
0010        Y(I)=.01*(1.+QC*RAN(IZUA,IZUB))
0011        A(I)=AFA
0012        B(I)=BFB
0013    140 CONTINUE
0014        A(KX)=AFA*QB
0015      2 ITOT=0
0016        WRITE (5,906)
0017    906 FORMAT (/,' RELATIVE ACTIVATOR CONCENTR.
            1AS FUNCTION OF CELL# AND TIME')
0018        IF(KY-KX.LE.14) WRITE (5,910) ((IK),IK=KX,KY)
0020    910 FORMAT (' A-MAXIMUM\    CELL NUMBER:',15I3)
0021      3 CALL PLOPP (A,AM,KX,KY,1,ITOT)    !    PRINTOUT
0022        IF (ITOT.GE.IC) GOTO 200          !  END OF THE CALCULATION
0024     40 DO 160 IP=1,IPR        !    START OF ITERATIONS
      C-----BOUNDARY CONDITIONS (IMPERMEABLE)
0025        A(KX-1)=A(KX)
0026        B(KX-1)=B(KX)
0027        A(KY+1)=A(KY)
0028        B(KY+1)=B(KY)
      C-----REACTIONS (CONCENTR.CHANGE PER ITERATION)
0029        DO 155 I=KX,KY        !    SEE EQUATION 3.2
0030        AQ=Y(I)*A(I)**2
0031        AD(I)=AQ/B(I)-TA*A(I)+DIA*(A(I-1)+A(I+1)-2.*A(I))+QA
0032        BD(I)=AQ-TB*B(I)+DIB*(B(I-1)+B(I+1)-2.*B(I))+QAA
0033    155 CONTINUE
      C-----ADDITION OF THE CHANGE TO THE EXISTING CONCENTRATIONS
0034        DO 160 I=KX,KY
0035        A(I)=A(I)+AD(I)
0036        B(I)=B(I)+BD(I)
0037    160 CONTINUE
0038        ITOT=ITOT+IPR
0039        GOTO 3
0040    200 WRITE(5, 909)
0041    909 FORMAT (/,' FINAL INHIBITOR DISTRIBUTION,MAX.AND %')
0042        CALL PLOPP (B,AM,KX,KY,2,ITOT)
0043        WRITE (5,908)
0044    908 FORMAT (/,' SOURCE DENSITY, MAXIMUM AND %')
0045        CALL PLOPP (Y,AM,KX,KY,4,ITOT)
0046        GOTO 150
0047        END
```

Program 17.1. Pattern formation in a one-dimensional array of cells.

```
  3600=IC         200=IPR   2=KX  10=KY    0=KZ    0=IA    0=IB    0=IZ
0.0200=DIA  0.4000=DIB  0.0100=TA  0.0200=TB  0.0001=QA  0.0000=QAA
1.0000=QB   0.0000=QBB  0.0200=QC  0.0000=QD  0.0000=QE  0.0000=QF
```

HOMOGENIOUS ACTIVATOR- AND INHIBITOR CON.   2.01000 2.02005

RELATIVE ACTIVATOR CONCENTR. AS FUNCTION OF CELL# AND TIME

```
A-MAXIMUM\    CELL NUMBER:  2   3   4   5   6   7   8   9 10
  2.0100A   AAAAAAAAA    100100100100100100100100100
  2.0321A   #######A#     98  98  98  98  99  99100100100
  2.0538A   #######A#     96  96  96  97  98  99  99100100
  2.0871A   #######A#     93  93  94  95  96  98  99100100
  2.1381A   ***##########A     88  89  90  91  93  96  98100100
  2.2109A   ****####A     82  83  84  87  90  94  97  99100
  2.3114A   $$$**###A     73  75  77  81  86  91  95  99100
  2.4481A   %%%$*#**A     62  64  68  73  80  87  93  98100
  2.6277A   ;++%$*##A     49  52  57  64  73  82  91  97100
  2.8495A   ::;+%$*#A     35  38  44  53  63  76  87  96100
  3.0945A   ,,:;+%*#A     22  25  31  41  53  68  83  94100
  3.3199A   ..,:;%$#A     13  15  21  30  44  60  78  92100
  3.4865A   ..,:+$#A       8  10  15  23  36  54  73  90100
  3.5895A   ..:;%*A        6   7  12  19  32  49  70  89100
  3.6461A   ..,;%*A        5   6  10  17  29  47  68  88100
  3.6751A   .,;%*A         4   6   9  16  28  45  67  88100
  3.6893A   .,;%*A         4   5   9  16  28  45  66  87100
  3.6962A   .,;%*A         4   5   9  16  27  44  66  87100
  3.6995A   .,;%*A         4   5   9  16  27  44  66  87100
```

FINAL INHIBITOR DISTRIBUTION,MAX.AND %

```
  2.9526B   :;;+%$*#A     39  41  45  52  60  71  83  93100
```

SOURCE DENSITY, MAXIMUM AND %

```
  0.0102Y   #######A#     98  98  98  98  99  99  99100  98
```

Simulation 17.1. Pattern formation in a one-dimensional array of cells. A stable graded concentration profile is generated out of random fluctuation (see Figs 3.2 and 4.1). Next page, top: the same parameters in a larger field leads to a symmetric distribution (see Fig. 4.1). Below: shorter range (higher decay rate) of the activator and inhibitor leads to a periodic pattern (see Fig. 4.8).

```
   4200=IC      300=IPR   2=KX  14=KY    0=KZ     0=IA     0=IB     0=IZ
0.0200=DIA  0.4000=DIB  0.0100=TA   0.0200=TB   0.0001=QA   0.0000=QAA
1.0000=QB   0.0000=QBB  0.0200=QC   0.0000=QD   0.0000=QE   0.0000=QF
```

HOMOGENIOUS ACTIVATOR- AND INHIBITOR CON. 2.01000 2.02005

RELATIVE ACTIVATOR CONCENTR. AS FUNCTION OF CELL# AND TIME

A-MAXIMUM\	CELL NUMBER:	2	3	4	5	6	7	8	9	10	11	12	13	14
2.0100A	AAAAAAAAAAAA	100	100	100	100	100	100	100	100	100	100	100	100	100
2.0325A	#####A#######	98	98	99	99	100	100	100	99	99	98	98	98	98
2.0706A	#####A#######	95	96	97	98	99	100	100	99	98	96	95	95	94
2.1237A	#####A#####**	90	92	94	96	99	100	100	99	96	94	92	90	89
2.2201A	****#A#*****	82	84	88	93	97	100	100	98	94	90	86	82	80
2.3805A	$$$*#*A#*%$$%	70	74	79	87	94	99	100	97	91	84	76	71	68
2.6202A	++%$*#A#*%%++	55	59	68	79	90	98	100	96	86	75	64	56	52
2.9098A	:;+%*#A#*%+;:	39	45	55	70	85	97	100	94	81	65	51	41	36
3.1550A	,:;%*A#$+;,,	28	34	45	62	81	95	100	92	76	57	41	30	24
3.2974A	,,:+$#A#$+:,.	22	28	40	57	78	95	100	91	73	52	35	24	19
3.3622A	.,:+$#A#$;:,.	19	25	37	55	77	94	100	91	71	50	32	22	16
3.3884A	.,:+$#A#$;:,.	18	24	36	54	76	94	100	91	70	49	31	21	16
3.3986A	.,:+$#A#$;:,.	18	24	35	54	76	94	100	91	70	48	31	20	15
3.4026A	.,:+$#A#$;:,.	18	23	35	54	76	94	100	91	70	48	31	20	15
3.4041A	.,:+$#A#$;:,.	18	23	35	53	75	94	100	91	70	48	31	20	15

FINAL INHIBITOR DISTRIBUTION,MAX.AND %

2.6364B	%%$*##A#*$$%%	65	68	74	82	91	98	100	96	88	78	70	65	62

SOURCE DENSITY, MAXIMUM AND %

0.0102Y	#####A#######	99	100	99	100	100	100	100	100	99	99	99	99	99

```
    300=IC      25=IPR   2=KX  15=KY    0=KZ     0=IA     0=IB     0=IZ
0.0100=DIA  0.4000=DIB  0.1000=TA   0.1500=TB   0.0001=QA   0.0000=QAA
1.0000=QB   0.0000=QBB  0.0200=QC   0.0000=QD   0.0000=QE   0.0000=QF
```

HOMOGENIOUS ACTIVATOR- AND INHIBITOR CON. 1.50100 0.15020

RELATIVE ACTIVATOR CONCENTR. AS FUNCTION OF CELL# AND TIME

A-MAXIMUM\	CELL NUMBER:	2	3	4	5	6	7	8	9	10	11	12	13	14	15
1.5010A	AAAAAAAAAAAA	100	100	100	100	100	100	100	100	100	100	100	100	100	100
1.5634A	##############A	93	99	98	93	95	99	97	94	97	97	96	93	93	100
1.7517A	$##$#*#$#*$#*$$A	76	96	92	75	81	97	87	77	87	92	88	76	74	100
2.2290A	;#*;+A%;%*$;:#	43	96	82	41	52	100	67	46	67	85	75	42	40	99
3.3008A	.*+..A:.:$;..$	14	84	52	12	19	100	32	14	34	78	49	11	13	78
3.6676A	*, .A. .#, %	9	85	24	5	11	100	14	4	14	93	26	4	8	69
3.6723A	*. .#. .A. %	10	89	11	2	10	99	10	2	11	100	13	2	7	68
3.6923A	* # .A. %	10	89	9	2	10	98	10	2	10	100	10	2	7	67
3.6953A	* # A. %	10	89	9	2	10	98	10	2	10	100	10	1	7	67
3.6949A	* # A. %	10	89	9	2	10	98	10	2	10	100	10	1	7	67
3.6950A	* # A. %	10	89	9	2	10	98	10	2	10	100	10	1	7	67
3.6950A	* # A. %	10	89	9	2	10	98	10	2	10	100	10	1	7	67
3.6950A	* # A. %	10	89	9	2	10	98	10	2	10	100	10	1	7	67

FINAL INHIBITOR DISTRIBUTION,MAX.AND %

0.3172B	%*++%#%+%A%;+$	66	89	60	52	64	98	65	55	66	100	63	48	51	73

SOURCE DENSITY, MAXIMUM AND %

0.0102Y	#####A#########	98	100	100	98	99	100	99	99	99	99	99	98	98	100

NEW CONSTANT # ^C

```
      C----PROGRAM LATERAL ACTIVATION
      C---(GENERATES STRIPES IN A TWO DIMENSIONAL FIELD)
      C---G1,G2(IX,IY) TWO SHORT-RANGING FEEDBACK LOOPS
      C---MADE LOCALLY EXCLUSIVE, E.G. BY COMMON REPRESSOR R(IX,IY)
      C---BOTH HELP EACH OTHER VIA THE LONG RANGING SUBSTANCES S1 OR S2
      C---DIFFERENT INTERACTION CAN BE CHOSEN BY IB
      C---IB=5 LATERAL INHIBITION MECHANISM FOR COMPARISON
      C
0001        DIMENSION G1(31,31),DG1(30),VZ(31,31)
           1,R(31,31),S1(31,31),DS1(30),DR(30),G2(31,31),DG2(30),
           2S2(31,31),DS2(30)
0002        COMMON/D/ IC,IPR,KX,KY,KZ,IA,IB,IZ,DIA,DIB,TA,TB,QA,
           1QAA,QB,QBB,QC,QD,QE,QF
0003        RR=RAN(0,0)
0004    150 CALL CONST
0005     50 WRITE (5,920)
0006        READ(5,931) IZK
0007        IF(IZK.EQ.0) GOTO 150
0009        GOTO (1,2,3,99),IZK
0010      1 GOTO (101,102,103,104,105),IB
      C---- CALCULATION OF THE SEMISTABLE EQUILIBRIA
      C-----FOR THE DIFFERENT TYPES OF REACTION
0011    101 WRITE (5, 901)    !      MULTIPLICATIVE HELP
0012        GINIT=TB/(2.*TA)
0013        RINIT=.02*GINIT**3/TB
0014        GOTO 125
0015    102 WRITE (5,902)     !      HELP BY SELFINHIBITON
0016        GINIT=TB/(2.*TA)
0017        RINIT=.02*GINIT/TB
0018        GOTO 125
0019    103 WRITE (5,903)     !      AUTOCATALYSIS REALIZED BY
0020        GINIT=SQRT(1.-QBB)  !    INHIBITION OF AN INHIBITON
0021        RINIT=1.
0022        GOTO 125
0023    104 WRITE (5,904)     !      STRONG ADDITIVE HELP
0024        GINIT=TB/(2.*TA)
0025        RINIT=.02*GINIT**2/TB
0026        GOTO 125
0027    105 WRITE (5,905)     !   AUTOCATALYSIS AND LATERAL INHIBITON
0028        GINIT=QD/TA
0029        SINIT=.01*GINIT**2/QD
0030        GOTO 126
0031    125 SINIT=GINIT
0032    126 DO 140 IY=1,KX          !   INITIAL CONDITION
0033        DO 140 IX=1,KY          ! = SEMESTABLE EQUILIBRUM
0034        G1(IX,IY)=GINIT
0035        G2(IX,IY)=GINIT
0036        R(IX,IY)=RINIT
0037        S1(IX,IY)=SINIT
0038        S2(IX,IY)=SINIT
0039        VZ(IX,IY)=(1.+RAN(IRAZ,IRBZ)*QC-.5*QC)*EXP(QF*(IX-1))
0040    140 CONTINUE
0041        JY=(KY-1)/2+1
0042        GOTO (141,142),IA    !   SELECTION OF THE INITIATION
```

Program 17.2. Lateral activation.

```
0043    141 WRITE (5,911)          !   HOMOGENIOUS,EXEPT RANDOM FLUCTUATION
0044        GOTO 2
0045    142 WRITE (5,912) KX,JY    !   LOCAL ADVANTAGE OF G1 AT IX,IY=
0046        G1(KX,JY)=1.1*GINIT
0047      2 ITOT=0
0048      3 WRITE (5,933)
       C---- PRINTOUT
0049        CALL XYPRIN (G1,AM,1,KX,1,KY,17,ITOT)
0050        IF (IB.EQ.5) GOTO 30
0052        WRITE (5,934)
0053        CALL XYPRIN (G2,AM,1,KX,1,KY,18,ITOT)
0054     30 IF (ITOT.GE.IC) GOTO 50
0056        DGC=1.-TA-4.*DIA       !   LOSS PER ITERATION BY DECAY AND
0057        DSC=1.-QD-4.*QE        !   DIFFUSION TO THE FOUR NEIGHBOURS
0058        DRC=1.-TB              !   (REPRESSOR IS NONDIFFUSIBLE)
0059      4 DO 160 IP=1,IPR        !   START OF THE ITERATIONS
       C------BOUNDARY CONDITION (IMPERMEABLE)
0060        DO 151 IX=1,KX
0061        G1(IX,KY+1)=G1(IX,KY)      !   LOWER BORDER
0062        G2(IX,KY+1)=G2(IX,KY)
0063        S1(IX,KY+1)=S1(IX,KY)
0064        S2(IX,KY+1)=S2(IX,KY)
0065        DG1(IX)=G1(IX,1)           !   UPPER BORDER
0066        DG2(IX)=G2(IX,1)
0067        DS1(IX)=S1(IX,1)
0068    151 DS2(IX)=S2(IX,1)
0069        DO 152 IY=1,KY
0070        G1(KX+1,IY)=G1(KX,IY)      !   RIGHT BORDER
0071        G2(KX+1,IY)=G2(KX,IY)
0072        S1(KX+1,IY)=S1(KX,IY)
0073    152 S2(KX+1,IY)=S2(KX,IY)
0074        DO 160 IY=1,KY
0075        G1L=G1(1,IY)               !   LEFT BORDER
0076        G2L=G2(1,IY)
0077        S1L=S1(1,IY)
0078        S2L=S2(1,IY)
       C------REACTION AND FEEDBACK------------
0079        DO 160 IX=1,KX
0080        GF1=G1(IX,IY)
0081        GF2=G2(IX,IY)
0082        RF=R(IX,IY)
0083        SF1=S1(IX,IY)
0084        SF2=S2(IX,IY)
0085        DDG1=DG1(IX)+G1(IX,IY+1)+G1L+G1(IX+1,IY)     !  GAIN BY DIFFUSIO
0086        DDG2=DG2(IX)+G2(IX,IY+1)+G2L+G2(IX+1,IY)     !  FROM THE FOUR
0087        DDS1=DS1(IX)+S1(IX,IY+1)+S1L+S1(IX+1,IY)     !  NEIGHBOURS
0088        DDS2=DS2(IX)+S2(IX,IY+1)+S2L+S2(IX+1,IY)
0089        GOTO (201,202,203,204,205),IB
0090        GOTO 155
0091    201 G1Q=.01*VZ(IX,IY)*GF1**2*SF2                 !  MULTIPLICATIVE HELP
0092        G2Q=.01*GF2**2*SF1                           !  EQ. 12.1
0093        G1(IX,IY)=GF1*DGC+DIA*DDG1+G1Q/RF+QA
0094        G2(IX,IY)=GF2*DGC+DIA*DDG2+G2Q/RF+QAA
0095        R(IX,IY)=RF*DRC+G1Q+G2Q
```

Program 17.2. (cont.)

```
0096          S1(IX,IY)=SF1*DSC+QE*DDS1+QD*GF1
0097          S2(IX,IY)=SF2*DSC+QE*DDS2+QD*GF2
0098          GOTO 155
0099      202 G1Q=.01*VZ(IX,IY)*GF1**2/SF1          ! HELP BY SELFINHIBITION
0100          G2Q=.01*GF2**2/SF2                     ! EQ. 12.2
0101          G1(IX,IY)=GF1*DGC+DIA*DDG1+G1Q/RF+QA
0102          G2(IX,IY)=GF2*DGC+DIA*DDG2+G2Q/RF+QAA
0103          R(IX,IY)=RF*DRC+G1Q+G2Q
0104          S1(IX,IY)=SF1*DSC+QE*DDS1+QD*GF1
0105          S2(IX,IY)=SF2*DSC+QE*DDS2+QD*GF2
0106          GOTO 155
0107      203 G1Q=SF2*TA*VZ(IX,IY)/(GF2**2+QBB)      ! AUTOCATALYSIS REALIZED BY
0108          G2Q=TA*SF1/(GF1**2+QBB)                ! INHIBITON OF INHIBITON
0109          G1(IX,IY)=GF1*DGC+DIA*DDG1+G1Q+QA      ! EQ. 12.3
0110          G2(IX,IY)=GF2*DGC+DIA*DDG2+G2Q+QAA
0111          S1(IX,IY)=SF1*DSC+QE*DDS1+QD*GF1
0112          S2(IX,IY)=SF2*DSC+QE*DDS2+QD*GF2
0113          GOTO 155
0114      204 GFS1=GF1+QA*SF2                        !STRONG ADDITIVE HELP
0115          G1Q=.01*VZ(IX,IY)*GFS1**2              ! EQ. 13.1
0116          GFS2=GF2+QA*SF1
0117          G2Q=.01*GFS2**2
0118          G1(IX,IY)=GF1*DGC+DIA*DDG1+G1Q/RF
0119          G2(IX,IY)=GF2*DGC+DIA*DDG2+G2Q/RF
0120          R(IX,IY)=RF*DRC+G1Q+G2Q
0121          S1(IX,IY)=SF1*DSC+QE*DDS1+QD*GF1
0122          S2(IX,IY)=SF2*DSC+QE*DDS2+QD*GF2
0123          GOTO 155
0124      205 G1Q=.01*VZ(IX,IY)*GF1**2               ! AUTOCATALYSIS AND
0125          G1(IX,IY)=GF1*DGC+DIA*DDG1+G1Q/SF1+QA  ! LATERAL INHIBITON
0126          S1(IX,IY)=SF1*DSC+QE*DDS1+G1Q          ! EQ.3.2 (TWO COMPONENTS)
0127      155 G1L=GF1
0128          DG1(IX)=GF1            ! PRESENT CELL (IX,IY) WITH ORIGINAL
0129          G2L=GF2               ! CONCENTRATION OF E.G.GF1 BECOMES LEFT
0130          DG2(IX)=GF2            ! NEIGHBOUR AT IX+1,IY AND
0131          S1L=SF1               ! UPPER NEIGHBOR AT IX,IY+1
0132          DS1(IX)=SF1
0133          S2L=SF2
0134          DS2(IX)=SF2
0135      160 CONTINUE
0136      163 ITOT=ITOT+IPR
0137      170 GOTO 3
0138       99 GINIT=0
0139      901 FORMAT (' REACTION TYPE - MULTIPLICATIVE HELP')
0140      902 FORMAT (' REACTION TYPE - HELP REALIZED BY SELFINHIBITION')
0141      903 FORMAT (' REACTION TYPE - INHIBITION OF AN INHIBITON')
0142      904 FORMAT (' REACTION TYPE - STRONG ADDITIVE HELP')
0143      905 FORMAT (' REACTION TYPE - AUTOCATAL. AND LATERAL INHIBITON')
0144      911 FORMAT (' HOMOGENIOUS,EXEPT RANDOM FLUCTUATION')
0145      912 FORMAT (' LOCAL ADVANTAGE OF G1 AT IX=',I2,' ,IY=',I2)
0146      920 FORMAT (' ? 0=NEW CONSTANTS, 1=START, 2=CONTINUATION, 3=PRINT')
0147      931 FORMAT (I6)
0148      933 FORMAT (/,'    G1-DISTRIBUTION')
0149      934 FORMAT ('    COMPLEMENTARY G2-DISTRIBUTION')
0150          END
```

Program 17.2. (cont.)

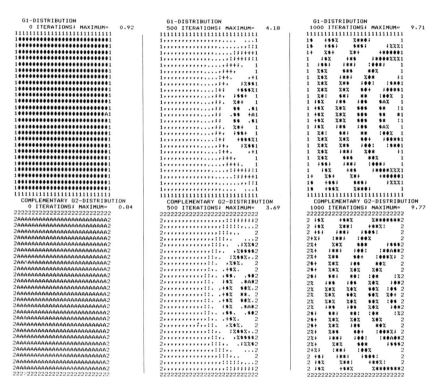

```
   1000=IC      500=IPR 25=KX   25=KY    0=KZ    2=IA    3=IB     0=IZ
  0.0050=DIA  0.0050=DIB  0.1000=TA   0.1000=TB  0.0010=QA   0.0010=QAA
  0.0300=QB   0.3000=QBB  0.0000=QC   0.0400=QD  0.2000=QE   0.0000=QF
? 0=NEW CONSTANTS, 1=START, 2=CONTINUATION, 3=PRINT
1
REACTION TYPE - INHIBITION OF AN INHIBITON
INITIAL CONDITION: LOCAL ADVANTAGE OF G1 AT IX=25 ,IY=13
```

Simulation 17.2. Pattern formation by lateral activation. This page: pattern is initiated by a small advantage of a single cell: stable stripes of high g_1 and g_2 concentrations are formed (see Fig. 12.2). Next page, top: same constants, initiation by random fluctuation, a stripe-like pattern with random orientation appears (see Fig. 12.2). Bottom next page: for comparison lateral inhibition (eq. 3.2) leads to a bristle-like pattern (see Fig. 4.9).

```
    500=IC        250=IPR  25=KX   25=KY    0=KZ    1=IA    1=IB     0=IZ
 0.0050=DIA  0.0050=DIB  0.0400=TA   0.0600=TB   0.0000=QA    0.0000=QAA
 0.0000=QB   0.0000=QBB  0.0200=QC   0.0400=QD   0.2000=QE    0.0000=QF
? 0=NEW CONSTANTS, 1=START, 2=CONTINUATION, 3=PRINT
1
REACTION TYPE - MULTIPLICATIVE HELP
INITIAL CONDITION: HOMOGENIOUS,EXEPT RANDOM FLUCTUATION
```

```
   G1-DISTRIBUTION                  G1-DISTRIBUTION                  G1-DISTRIBUTION
    0 ITERATIONS; MAXIMUM=  0.75     250 ITERATIONS; MAXIMUM=  1.53    500 ITERATIONS; MAXIMUM=   1.55
 1111111111111111111111111111    1111111111111111111111111111    1111111111111111111111111111
 1AAAAAAAAAAAAAAAAAAAAAAAAAAA1    (ASCII distribution art)        (ASCII distribution art)
 ...
   COMPLEMENTARY G2-DISTRIBUTION    COMPLEMENTARY G2-DISTRIBUTION    COMPLEMENTARY G2-DISTRIBUTION
    0 ITERATIONS; MAXIMUM=  0.75     250 ITERATIONS; MAXIMUM=  1.57    500 ITERATIONS; MAXIMUM=   1.44
 2222222222222222222222222222    2222222222222222222222222222    2222222222222222222222222222
 ...
```

```
   1000=IC       500=IPR  25=KX   25=KY    0=KZ    1=IA    5=IB     0=IZ
 0.0100=DIA  0.0000=DIB  0.0300=TA   0.0000=TB   0.0000=QA    0.0000=QAA
 0.0000=QB   0.0000=QBB  0.0200=QC   0.0400=QD   0.2000=QE    0.0000=QF
? 0=NEW CONSTANTS, 1=START, 2=CONTINUATION, 3=PRINT
1
REACTION TYPE - AUTOCATAL. AND LATERAL INHIBITON
INITIAL CONDITION: HOMOGENIOUS,EXEPT RANDOM FLUCTUATION
```

```
   G1-DISTRIBUTION                  G1-DISTRIBUTION                  G1-DISTRIBUTION
    0 ITERATIONS; MAXIMUM=  1.33     500 ITERATIONS; MAXIMUM=  4.04   1000 ITERATIONS; MAXIMUM=   6.04
 1111111111111111111111111111    1111111111111111111111111111    1111111111111111111111111111
 1AAAAAAAAAAAAAAAAAAAAAAAAAAA1    (ASCII distribution art)        (ASCII distribution art)
 ...
 1111111111111111111111111111    1111111111111111111111111111    1111111111111111111111111111
```

```
      C-----PROGRAM COMPARTMENT (CREATES A SUPERPOSITION OF A PERIODIC
      C-----AND A SEQUENTIAL PATTERN (COMPARTMENTS AND GENE ACTIVITIES)
      C-----G(I,IG): GENE ACTIVATOR   I=POSITION, IG=GENE#
      C-----GEREPR(I): COMMON REPRESSOR OF THE GENE ACTIVATORS
      C-----DUE TO HIERACHY X(IG), GEREPR ACTS AS POS.VALUE
      C-----GTRANS(I) INDUCES TRANSITION FROM GENE #I TO #I+1
      C-----A(I),P(I):ANTERIOR AND POSTERIOR DETERMINATION
      C-----RAP(I):COMMON REPRESSOR OF A(I) AND P(I)
      C-----SA(I),SP(I): LONG RANGE MUTUAL HELP OF THE TWO COMPARTMENTS
      C-----POSINF(I): POSITIONAL INFORMATION (MORPHOGEN GRADIENT)
      C
0001        COMMON/D/ IC,IPR,KX,KY,KZ,IA,IB,IZ,DIA,DIB,TA,TB,QA,
           1QAA,QB,QBB,QC,QD,QE,QF
0002        DIMENSION POSINF(30),A(32),RAP(32),P(31),SA(31),SP(30)
           1,G(31,10),GTRANS(31,10),GEREPR(31),X(31)
           2,TY(20),TYA(30),TYP(30)
0003        DATA TY/' ',',',',',',',':',';',';','+','%','$','*','#','A','B',
           1'C','Y','X','N','P','S','[',']'/
0004    150 CALL CONST(11)
0005     50 WRITE (5,920)
0006    920 FORMAT (' ?,0=NEW CONST.,1=START,2=CONTINUATION,3=PRINT,4=END')
0007        READ(5,960) IZK
0008        IF(IZK.EQ.0) GOTO 150
0010        GOTO (1,2,210,99),IZK
      C-----INITIAL CONDITION
0011      1 AINIT=TB/(2.*TA)       ! SEMISTABLE EQUILIBRUM
0012        RAPIN=.02*AINIT/TB
0013        DO 108 I=1,30
0014        X(I)=.01*EXP(-.3*FLOAT (I-1))    ! HIERACHY OF THE GENES
0015        POSINF(I)=QF*EXP(QC*FLOAT(I-1))  ! POSITIONAL INFORMATION
0016        IF(QF.LT.0.) POSINF(I)=          ! P.I.FOR BICAUDAL
           1-QF*EXP(QC*FLOAT(I-1))-QF*EXP(QC*FLOAT(KX-I))
0018        P(I)=AINIT                       ! INITIALLY , ALL CELLS
0019        A(I)=0.                          ! ARE POSTERIOR
0020        RAP(I)=RAPIN
0021        SA(I)=0.1
0022        SP(I)=AINIT
0023        G(I,1)=1.5          ! ONLY GENE #1 IS 'ON'
0024        GEREPR(I)=.01*.3/.2**2
0025        TYA(I)=TY(1)
0026        DO 108 IG=2,10
0027        G(I,IG)=0.          ! OTHER GENES ARE OFF
0028        GTRANS(I,IG)=0.
0029    108 CONTINUE
0030        WRITE (5,905)
0031    905 FORMAT (/,' POSITIONAL INFORMATION IN THE CELL 1.....KX')
0032        WRITE (5,906) (POSINF(IL),IL=1,KX)
0033        WRITE (5,911)
0034    911 FORMAT (/,' GENE-ACTIVATOR AS FUNTION OF POSITION (X) AND GENE #
0035        WRITE (5,913)
0036    913-FORMAT (' INITIALLY, ONLY GENE #1 IS ACTIVE IN EVERY CELL',/)
0037        ITOT=0
0038        CALL XYPRIN (G,AM,1,KX,1,KY,10,ITOT) !PLOT OF GENE-ACTIVATORS
0039        IF (IA.NE.2) GOTO 2
```

Program 17.3. Compartment.

```
0041          A(1)=AINIT           !   IN PROGRESS-ZONE MODEL,
0042          P(1)=0.1             !   LEFTMOST CELL IS ANTERIOR
0043       2 ITOT=0
0044          IGROW=0
0045          DAC=1.-TA
0046          DBC=1.-TB            ! LOSS PER ITERATION BY DECAY
0047          DSA=1.-QD-2.*DIA     ! (AND BY DIFFUSION)
0048          DSP=1.-QD-2.*DIB
0049          WRITE  (5,907)
C-----PREPARATION OF THE PRINTOUT OF THE ANTERIOR-POSTERIOR DISTRIB.
0050       3 DO 20 I=1,KX
0051          NA=A(I)*5.+1
0052          TYA(I)=TY(NA)
0053          NP=P(I)*5.+1
0054      20 TYP(I)=TY(NP)
0055          TYA(KX+1)=TY(20)
0056          WRITE (5,910) ITOT,TY(19),(TYA(IK),IK=1,30)
             1,TY(19),(TYP(IM),IM=1,KX),TY(20)
0057          IF (ITOT.GE.IC) GOTO 210     !    END OF CALCULATION IS REACHED
0059          IF(IZ.GT.0.AND.IGROW.GE.IZ) GOTO 500   ! USED IF FIELD IS GROWING
0061      30 DO 100 ICC=1,IPR       ! BEGIN OF THE ITERATION
0062          SA1=SA(1)
0063          SP1=SP(1)
0064          SA(KX+1)=SA(KX)
0065          SP(KX+1)=SP(KX)
C-----INTERACTION OF "ANTERIOR" AND "POSTERIOR"
C-----LEADING TO OSCILLATIONS AND STRIPES
0066          DO 90 I=1,KX
0067          AF=A(I)
0068          RF=RAP(I)
0069          PF=P(I)
0070          SAF=SA(I)
0071          SPF=SP(I)
0072          THRESH=QB/(POSINF(I)*GEREPR(I))   ! DETERMINES THRESCHOLD FOR
0073          AQF=.01*AF**2/(SAF+THRESH)        ! A FURTHER OSCILLATION
0074          PQF=.01*PF**2/SPF      !  MUTUAL HELP OF ANT. AND POST. BY SELF-
0075          A(I)=AF*DAC+AQF/RF+QA    !  INHIBITION, SEE EQUATION 12.2
0076          P(I)=PF*DAC+PQF/RF+QA
0077          RAP(I)=RF*DBC+AQF+PQF
0078          SA(I)=SAF*DSA+QD*AF+DIA*(SA1+SA(I+1))
0079          SP(I)=SPF*DSP+QD*PF+DIB*(SP1+SP(I+1))
0080          GOTO 80
0081      80 SA1=SAF   !  ORIGINAL CONCENTRATION AT I IS USED AS CONCENTRATION
0082          SP1=SPF   !  OF THE LEFT NEIGHBOUR AT POSITION I+1
C-----MAINTENANCE OF THE GENE ACTIVITY AND TRANSITION TO THE NEXT GENE
0083          RESUM=0.              !SEE EQUATION 11.3 AND 11.4
0084          GREPR=GEREPR(I)
0085          DO 87 IG=1,KY
0086          GF=G(I,IG)
0087          TRANS=0. ! ACTION OF THE TRANS-MOLECULE IS BLOCKED IN THE POSTE-
0088          IF (IG.GE.2) TRANS=QAA*GTRANS(I,IG-1)/PF   !  RIOR COMPARTMENT
0090          GQ=X(IG)*(GF+TRANS)**2
0091          G(I,IG)=.8*GF+GQ/GREPR
0092          RESUM=RESUM+GQ
```

Program 17.3. (cont.)

```
          C---- MOLECULES INDUCING TRANSITION ARE PRODUCED ONLY
          C-----IN THE STATE "POSTERIOR"
0093         GTRANS(I,IG)=GTRANS(I,IG)*(1.-QBB)+QBB*PF*G(I,IG)
0094      87 CONTINUE
0095         GEREPR(I)=.7*GREPR+RESUM   !LOSS BY DECAY OF GENE REPRESSOR
0096      90 CONTINUE                   ! + SUM OF THE NEW PRODUCTION
0097     100 CONTINUE
0098       4 ITOT=ITOT+IPR
0099         IGROW=IGROW+1
0100         GOTO 3
0101     210 WRITE (5,911)
0102         CALL XYPRIN (G,AM,1,KX,1,KY,10,ITOT) !PLOT OF GENE-ACTIVATORS
0103         WRITE (5,917)
0104         WRITE (5,918) TY(11),(TYA(IK),IK=1,KX),TY(11)
0105         WRITE (5,918) TY(17),(TYP(IK),IK=1,KX),TY(17)
0106         WRITE (5,912)
0107     912 FORMAT (' MOLECULES INDUCING TRANSITION ARE PRODUCED ONLY',/,
                1' IN THE POSTERIOR COMPARTMENT OF EACH SEGMENT')
0108         CALL XYPRIN (GTRANS,AM,1,KX,1,KY,6,ITOT)
0109         GOTO 50
0110     500 KX=KX+1        ! GROWTH OF THE FIELD AT THE RIGHT MARGIN
0111         IGROW=0
0112         A(KX)=A(KX-1)
0113         P(KX)=P(KX-1)
0114         RAP(KX)=RAP(KX-1)
0115         SA(KX)=SA(KX-1)
0116         SP(KX)=SP(KX-1)
0117         GEREPR(KX)=GEREPR(KX-1)
0118         DO 501 IG=1,KY
0119         GTRANS(KX,IG)=GTRANS(KX-1,IG)
0120     501 G(KX,IG)=G(KX-1,IG)
0121         GOTO 30
0122      99 I=0
0123     906 FORMAT (13F5.2)
0124     907 FORMAT (/,' "ANTERIOR" (LEFT) AND COMPLEMENTARY "POSTERIOR" AS',
                1/,' FUNCTION OF POSTION (X) AND TIME (ITERATIONS)')
0125     910 FORMAT (1X,I6,2X,70A1)
0126     917 FORMAT (' DISTRIBUTION OF HIGH A AND HIGH P')
0127     918 FORMAT (1X,30A1)
0128     960 FORMAT (I6)
0129         END
```

Simulation 17.3. Next three pages: Interpretation of positional information: formation of a periodic pattern of anterior and posterior compartmental specifications and, in register, the activation and particular control genes. Normal monotonic gradient, symmetric gradient (see Fig. 14.5) and pattern formation in an area of marginal growth (see Fig. 14.8).

```
    1800=IC       60=IPR 25=KX    6=KY    0=KZ    0=IA    0=IB    0=IZ
  0.0010=DIA   0.0020=DIB   0.1000=TA    0.1500=TB    0.0012=QA    0.1000=QAA
  0.0030=QB    0.0100=QBB   0.0650=QC    0.0500=QD    0.0000=QE    2.1000=QF
?,0=NEW CONST.,1=START,2=CONTINUATION,3=PRINT,4=END
1


POSITIONAL INFORMATION IN THE CELL 1.....KX
2.10 2.24 2.39 2.55 2.72 2.91 3.10 3.31 3.53 3.77 4.02 4.29 4.58
4.89 5.22 5.57 5.94 6.34 6.77 7.22 7.71 8.22 8.78 9.36 9.99

GENE-ACTIVATOR AS FUNTION OF POSITION (X) AND GENE #
INITIALLY, ONLY GENE #1 IS ACTIVE IN EVERY CELL

    0 ITERATIONS; MAXIMUM=   1.50
*******************************
#AAAAAAAAAAAAAAAAAAAAAAAAAAAA#
#                            #
#                            #
#                            #
#                            #
#                            #
*******************************
"ANTERIOR" (LEFT) AND COMPLEMENTARY "POSTERIOR" AS
FUNCTION OF POSTION (X) AND TIME (ITERATIONS)
    0  [                          ]  [::::::::::::::::::::::::::::::]
   60  [                          ]  [****************************]
  120  [       .,;%%%++++++++++++] [$$$$$$$%%+;;,...             ]
  180  [     ;+%$$$%%%%$$$$$$$$$$] [$$$%;.                      ]
  240  [  %$$$$$$$$$$$%%+;;:;,,,] [$$$.      .,:;%%%%%%%]
  300  [  $$$:.                ] [$$$    %+;+%$***$%%%%%%%]
  360  [  $%                   ] [$$$   *%$$*$$$$$$$$$$$$$$$]
  420  [  $:          .,+%%%++++] [$$$ +$$$$$$$$$$%+;;,..     ]
  480  [  $        .%+%$$$$%%%$$$] [$$$ $$$$$$+,           ]
  540  [  $      :%$$$$$$$$$$%%+;] [$$$ $$$$+        .:+%]
  600  [  $      $$$$%,.        ] [$$$ $$$$   ,%+;+%***$$$%]
  660  [  $     $$            ] [$$$ $$$$  +$$%$$*$$$$$$$$$]
  720  [  $     $        ,+%%] [$$$ $$$$ %$$$$$$$$$$%%;;,]
  780  [  $     $      .+%+$*$$%%] [$$$ $$$$ $$$$$$%:.      ]
  840  [  $     $    ,%$$$$$$$$$] [$$$ $$$$ $$$$+          ]
  900  [  $     $    $$$$$;.   ] [$$$ $$$$ $$$$   %%;+%**]
  960  [  $     $    $$.     ] [$$$ $$$$ $$$$   %%*%%$$$$$]
 1020  [  $     $    $      ] [$$$ $$$$ $$$$ +%$$$$$$$$$$]
 1080  [  $     $    $   .+++$] [$$$ $$$$ $$$$ $$$$$$%:.   ]
 1140  [  $     $    $  :+$$$$$] [$$$ $$$$ $$$$ $$$$+      ]
 1200  [  $     $    $ ,$$$$%:.] [$$$ $$$$ $$$$ $$$+   .%+]
 1260  [  $     $    $ $$$.   ] [$$$ $$$$ $$$$ $$$   +%*%%]
 1320  [  $     $    $ $$    ] [$$$ $$$$ $$$$ $$$   $$$$$$]
 1380  [  $     $    $ $$    ] [$$$ $$$$ $$$$ $$$   $$$$$$]
 1440  [  $     $    $ $$   +%] [$$$ $$$$ $$$$ $$$   $$$$: ]
 1500  [  $     $    $ $$  %$$] [$$$ $$$$ $$$$ $$$   $$$.  ]
 1560  [  $     $    $ $$  $$;] [$$$ $$$$ $$$$ $$$   $$$  %]
 1620  [  $     $    $ $$  $$ ] [$$$ $$$$ $$$$ $$$   $$$  %]
 1680  [  $     $    $ $$  $$ ] [$$$ $$$$ $$$$ $$$   $$$  $]
 1740  [  $     $    $ $$  $$ ] [$$$ $$$$ $$$$ $$$   $$$  $]
 1800  [  $     $    $ $$  $$ ] [$$$ $$$$ $$$$ $$$   $$$  $]

GENE-ACTIVATOR AS FUNTION OF POSITION (X) AND GENE #
 1800 ITERATIONS; MAXIMUM=   1.50
*******************************
#***#                        #
#  #####                     #
#     #####                  #
#        ####                #
#           #####            #
#               ##A#         #
*******************************
DISTRIBUTION OF HIGH A AND HIGH P
A   $   $    $   $$   $$ A
P$$$ $$$$ $$$$ $$$  $$$  $P
MOLECULES INDUCING TRANSITION ARE PRODUCED ONLY
IN THE POSTERIOR COMPARTMENT OF EACH SEGMENT
 1800 ITERATIONS; MAXIMUM=   2.24
++++++++++++++++++++++++++++
+###                        +
+  A###                     +
+     ####                  +
+        ###                +
+           ###             +
+              #+           +
++++++++++++++++++++++++++++
```

```
    1800=IC        60=IPR 25=KX   6=KY   0=KZ    0=IA    0=IB    0=IZ
  0.0010=DIA  0.0020=DIB   0.1000=TA    0.1500=TB     0.0012=QA   0.1000=QAA
  0.0030=QB   0.0100=QBB   0.1000=QC    0.0500=QD     0.0000=QE  -0.8200=QF
?,0=NEW CONST.,1=START,2=CONTINUATION,3=PRINT,4=END
1

POSITIONAL INFORMATION IN THE CELL 1.....KX
9.86 9.09 8.40 7.80 7.28 6.83 6.45 6.14 5.89 5.69 5.55 5.47 5.44
5.47 5.55 5.69 5.89 6.14 6.45 6.83 7.28 7.80 8.40 9.09 9.86

GENE-ACTIVATOR AS FUNTION OF POSITION (X) AND GENE #
INITIALLY, ONLY GENE #1 IS ACTIVE IN EVERY CELL

     0 ITERATIONS; MAXIMUM=   1.50
******************************
#AAAAAAAAAAAAAAAAAAAAAAAAAA#
#                          #
#                          #
#                          #
#                          #
#                          #
******************************
'ANTERIOR' (LEFT) AND COMPLEMENTARY 'POSTERIOR' AS
FUNCTION OF POSTION (X) AND TIME (ITERATIONS)
     0  [                      ]  [:::::::::::::::::::::::::::]
    60  [                      ]  [**************************]
   120  [+++++++++++++++++++++++]  [        .......          ]
   180  [$$$$$$$$$$$$$$$$$$$$$$$]  [                         ]
   240  [,,::;;++%%%%%%%%++;;;,,]  [%%%%%%+;;,,,,,,;;+%%%%%%]
   300  [                      ]  [%%%%%%+$$$$$$$$$$$+%%%%%%]
   360  [                      ]  [$$$$$$$$$$$$$$$$$$$$$$$$$]
   420  [++++%%+;,,......,;+%%++++]  [  .,,;:++%%%%%%++;,,..  ]
   480  [$$$%%%$$$$$$*$*$$$$%%%$$$]  [                         ]
   540  [;+%$$$$$$$$$$$$$$$$%+;]  [%;,.            .,;%]
   600  [                      ]  [%$***%$%+;;;;;;;+%$***$%]
   660  [                      ]  [$$$$$$$$$$$$%$$$$$$$$$]
   720  [%%,.            .:%%]  [,:+%$$$$$$$$$$$$$$$%+:,]
   780  [%%$*$*$*$++%+,.  .,,+%++$*%$%%]  [     ,:+%%%%%+:,      ]
   840  [$$$$$$$%$$$$$$$$$%$$$$$$$]  [                         ]
   900  [   .:%$$$$$$$$$$$%:.   ]  [*$++;+%,        ,%+;+$*]
   960  [                      ]  [$$$$$%%$*$$+;;+;;+$*%$%$$$$$]
  1020  [                      ]  [$$$$$$$$$$*%$$$$$*%$$$$$$$$$]
  1080  [%+%.              .%+%]  [ ,%$$$$$$$$$$$$$$%$%, ]
  1140  [$$$$%+           +%$$$$]  [   :%$$$$$$$$$$$$%:    ]
  1200  [,+$$$$%          %$$$$+,]  [%:   %$$$$$$$$$%    :%]
  1260  [  ,%$$          $$%,  ]  [%$$+.  %$$$$$$$$%  .+$%]
  1320  [    :$          $:    ]  [$$$$$+ $$$$$$$$$$ +$$$$$]
  1380  [     $          $     ]  [$$$$$$ $$$$$$$$$$ $$$$$$]
  1440  [+.   $          $   .+]  [ %$$$$ $$$$$$$$$$ $$$$% ]
  1500  [$%   $          $   %$]  [  $$$$ $$$$$$$$$$ $$$$  ]
  1560  [$$   $          $   $$]  [  $$$$ $$$$$$$$$$ $$$$  ]
  1620  [ $   $          $   $ ]  [$ $$$$ $$$$$$$$$$ $$$$ $]
  1680  [ $   $          $   $ ]  [$ $$$$ $$$$$$$$$$ $$$$ $]
  1740  [ $   $          $   $ ]  [$ $$$$ $$$$$$$$$$ $$$$ $]
  1800  [ $   $          $   $ ]  [$ $$$$ $$$$$$$$$$ $$$$ $]

GENE-ACTIVATOR AS FUNTION OF POSITION (X) AND GENE #
  1800 ITERATIONS; MAXIMUM=   1.50
******************************
#                          #
#                          #
#                          #
#                          #
#       ************       #
#  *****         *****  #
#A#                  #A#
******************************
DISTRIBUTION OF HIGH A AND HIGH P
A $         $         $  $ A
P$ $$$$ $$$$$$$$$$$$ $$$$ $P
MOLECULES INDUCING TRANSITION ARE PRODUCED ONLY
IN THE POSTERIOR COMPARTMENT OF EACH SEGMENT
  1800 ITERATIONS; MAXIMUM=   2.24
++++++++++++++++++++++++++++++
+                          +
+                          +
+                          +
+                          +
+       A*********A        +
+  ****         ****  +
+*                  *+
++++++++++++++++++++++++++++++
```

Computer programs

205

```
   6000=IC        180=IPR    3=KX    4=KY    0=KZ    2=IA    0=IB    2=IZ
0.1000=DIA   0.1000=DIB   0.0500=TA   0.0700=TB   0.0010=QA   0.1000=QAA
0.0000=QB    0.0100=QBB   0.0000=QC   0.1000=QD   0.0000=QE   1.0000=QF
?,0=NEW CONST.,1=START,2=CONTINUATION,3=PRINT,4=END
1


POSITIONAL INFORMATION IN THE CELL 1.....KX
1.00 1.00 1.00

GENE-ACTIVATOR AS FUNTION OF POSITION (X) AND GENE #
INITIALLY, ONLY GENE #1 IS ACTIVE IN EVERY CELL

      0 ITERATIONS; MAXIMUM=   1.50
#####                      1   2   3
#AAA#                    100100100
#   #                      0   0   0
#   #                      0   0   0
#   #                      0   0   0
#####

'ANTERIOR' (LEFT) AND COMPLEMENTARY 'POSTERIOR' AS
FUNCTION OF POSTION (X) AND TIME (ITERATIONS)
       0  [:    ]                      [ ::]
     180  [$    ]                      [ $$]
     360  [$    ]                      [ $$]
     540  [$    ]                      [ $$$]
     720  [$    ]                      [ $$$]
     900  [$     ]                     [ $$$%]
    1080  [$    $]                     [ $$$ ]
    1260  [$    $$]                    [ $$$   ]
    1440  [$    $$]                    [ $$$   ]
    1620  [$    $$$]                   [ $$$    ]
    1800  [$    $$$]                   [ $$$    ]
    1980  [$    $$$%]                  [ $$$     ]
    2160  [$    $$$ ]                  [ $$$    %]
    2340  [$    $$$  ]                 [ $$$  $$]
    2520  [$    $$$  ]                 [ $$$  $$]
    2700  [$    $$$   ]                [ $$$  $$$]
    2880  [$    $$$   ]                [ $$$  $$$]
    3060  [$    $$$    ]               [ $$$  $$$%]
    3240  [$    $$$   %]               [ $$$  $$$ ]
    3420  [$    $$$   $$]              [ $$$  $$$  ]
    3600  [$    $$$   $$]              [ $$$  $$$  ]
    3780  [$    $$$   $$$]             [ $$$  $$$   ]
    3960  [$    $$$   $$$]             [ $$$  $$$   ]
    4140  [$    $$$   $$$%]            [ $$$  $$$    ]
    4320  [$    $$$   $$$ ]            [ $$$  $$$   %]
    4500  [$    $$$   $$$  ]           [ $$$  $$$  $$]
    4680  [$    $$$   $$$  ]           [ $$$  $$$  $$]
    4860  [$    $$$   $$$   ]          [ $$$  $$$  $$$]
    5040  [$    $$$   $$$   ]          [ $$$  $$$  $$$]
    5220  [$    $$$   $$$    ]         [ $$$  $$$  $$$%]
    5400  [$    $$$   $$$   %]         [ $$$  $$$  $$$ ]
    5580  [$    $$$   $$$   $$]        [ $$$  $$$  $$$   ]
    5760  [$    $$$   $$$   $$]        [ $$$  $$$  $$$   ]
    5940  [$    $$$   $$$   $$$]       [ $$$  $$$  $$$    ]
    6120  [$    $$$   $$$   $$$]       [ $$$  $$$  $$$    ]

GENE-ACTIVATOR AS FUNTION OF POSITION (X) AND GENE #
   6120 ITERATIONS; MAXIMUM=   1.50
######################
#####            #
#    ######      #
#        ######  #
#            ##A#
######################
DISTRIBUTION OF HIGH A AND HIGH P
A$   $$$   $$$   $$$A
P $$$   $$$   $$$  P
MOLECULES INDUCING TRANSITION ARE PRODUCED ONLY
IN THE POSTERIOR COMPARTMENT OF EACH SEGMENT
   6120 ITERATIONS; MAXIMUM=   2.11
++++++++++++++++++++
+ ###            +
+       ###      +
+           ##A  +
+                +
++++++++++++++++++++
```

```
      C PROGRAM BRANCH (GENERATES A NETLIKE STRUCTURE)
      C-----A(IX,IY)=ACTIVATOR, B(IX,IY)=INHIBITOR
      C-----S(IX,IY)=DEPLETED SUBSTANCE Y(IX,IY)=DIFFERENTIATION
0001        COMMON/D/ IC,IPR,KX,KY,KZ,IA,IB,IZ,DIA,DIB,TA,TB,QA,
           1 QAA,QB,QBB,QC,QD,QE,QF
0002        DIMENSION A(31,31),DA(30,30),VZ(31,31),Y(31,31)
           1 ,B(31,31),S(31,31),DB(30,30),DS(30,30),TY(6),TYP(31),TYS(31)
0003        DATA TY/' ',':','.','A','*','!'/,TYS/31*'-'/
0004        RRN=RAN(0,0)
0005    150 CALL CONST
0006     50 WRITE (5,920)
0007        READ(5,931) IZK
0008        IF(IZK.EQ.0) GOTO 150
0010        GOTO (1,2,3,99),IZK
      C-----INITIAL CONDITION
0011      1 DO 140 IY=1,30
0012        DO 140 IX=1,30
0013        A(IX,IY)=.001           !      ACTIVATION OF THE CELL IS LOW
0014        VZ(IX,IY)=QAA*(1.+QC*RAN(IRANZ,IRBNZ))
0015        Y(IX,IY)=0.001          !      CELLS ARE NOT DIFFRENTIATED
0016        S(IX,IY)=1.
0017        B(IX,IY)=.01
0018    140 CONTINUE
0019        Y(3,3)=1.
0020        WRITE (5,900)
      C-----INITIAL CONDITION: ONE DIFFERENTIATED CELL
      C-----.... = S>.85;    AAA = A>.85;    *** = Y>.85
0021      2 ITOT=0
0022      3 WRITE (5,933) ITOT
      C-----PRINTOUT
0023        TYP(KX+1)=TY(6)
0024        TYP(1)=TY(6)
0025        WRITE (5,934) (TYS(IL),IL=1,KX+1)
0026        DO 170 IY=2,KY
0027        DO 171 IX=2,KX
0028        TYP(IX)=TY(1)
0029        IF (S(IX,IY).GT.0.85) TYP(IX)=TY(3)
0031        IF (A(IX,IY).GT.0.85) TYP(IX)=TY(4)
0033        IF (Y(IX,IY).GT.0.85) TYP(IX)=TY(5)
0035    171 CONTINUE
0036    170 WRITE (5,934) (TYP(IL),IL=1,KX+1)
0037        WRITE (5,934) (TYS(IL),IL=1,KX+1)
0038        IF (ITOT.GE.IC)    GOTO 150.
0040        DAC=1.-TA-4.*DIA      !     LOSS BY DECAY AND DIFFUSION
0041        DBC=1.-TB-4.*DIB      !      IN EACH ITERATION
0042        DSC=1.-QBB-4.*QE
0043        DYC=1.-.1
0044     40 DO 160 IP=1,IPR
      C-----BOUNDARY CONDITION (IMPERMEABLE)
0045        DO 151 IX=2,KX
0046        A(IX,KY+1)=A(IX,KY)
0047        B(IX,KY+1)=B(IX,KY)
0048        S(IX,KY+1)=S(IX,KY)
0049        A(IX,1)=A(IX,2)
```

Program 17.4. Branch (generates a net-like structure).

```
0050          S(IX,1)=S(IX,2)
0051      151 B(IX,1)=B(IX,2)
0052          DO 152 IY=2,KY
0053          A(KX+1,IY)=A(KX,IY)
0054          A(1,IY)=A(2,IY)
0055          S(KX+1,IY)=S(KX,IY)
0056          S(1,IY)=S(2,IY)
0057          B(KX+1,IY)=B(KX,IY)
0058      152 B(1,IY)=B(2,IY)
      C-----DIFFUSION
0059          DO 10 IX=2,KX
0060          DO 10 IY=2,KY
0061          DA(IX,IY)=A(IX,IY-1)+A(IX,IY+1)+A(IX-1,IY)+A(IX+1,IY)
0062          DS(IX,IY)=S(IX,IY-1)+S(IX,IY+1)+S(IX-1,IY)+S(IX+1,IY)
0063          DB(IX,IY)=B(IX,IY-1)+B(IX,IY+1)+B(IX-1,IY)+B(IX+1,IY)
0064       10 CONTINUE
      C-----REACTION
0065          DO 160 IX=2,KX
0066          DO 160 IY=2,KY
0067          AF=A(IX,IY)
0068          SF=S(IX,IY)
0069          BF=B(IX,IY)
0070          YF=Y(IX,IY)
0071          AQ=AF**2*SF*VZ(IX,IY)          !  SEE EQ. 15.1
0072          YQ=YF**2
0073          BQ=AQ/BF+QA*YF
0074          A(IX,IY)=AF*DAC+DIA*DA(IX,IY)+BQ              ! (EQ. 15.1a)
0075          B(IX,IY)=BF*DBC+DIB*DB(IX,IY)+AQ+QD*YF+.00002  ! (EQ. 15.1b)
0076          S(IX,IY)=SF*DSC+QE*DS(IX,IY)-QF*SF*YF+QBB     ! (EQ. 15.1d)
0077          Y(IX,IY)=YF*DYC+YQ/(1.+9.*YQ)+QB*AF+.0001     ! (EQ. 15.1c)
0078      160 CONTINUE
0079      163 ITOT=ITOT+IPR
0080          GOTO 3
0081       99 IC=0
0082      900 FORMAT (' INITIAL CONDITION: ONE DIFFERENTIATED CELL (#)',/
          1,' .... = S>.85;    AAA = A>.85;    ### = Y>.85',/)
0083      920 FORMAT (' ? 0=NEW CONSTANTS,1=START, 2=CONTINUE, 3=PRINT, 4=END')
0084      931 FORMAT (I6)
0085      933 FORMAT (I6,' ITERATIONS')
0086      934 FORMAT (' ',33A1)
0087          END
```

Program 17.4. (cont.)

```
10000=IC    2000=IPR 18=KX   18=KY    0=KZ    0=IA    0=IB    0=IZ
0.0200=DIA  0.2000=DIB  0.1200=TA   0.0400=TB    0.0400=QA    0.0040=QAA
0.0014=QB   0.0200=QBB  0.0300=QC    0.0003=QD    0.0600=QE    0.2000=QF
? 0=NEW CONSTANTS,1=START, 2=CONTINUE, 3=PRINT, 4=END
1
```

INITIAL CONDITION: ONE DIFFERENTIATED CELL (#)
.... = S>.85;   AAA = A>.85;   ### = Y>.85

```
       0 ITERATIONS                       6000 ITERATIONS
  --------------------            --------------------
  !..................!            ! A        A        !
  !.#................!            ! #        #        !
  !..................!            ! #        #        !
  !..................!            !  #  ############# !
  !..................!            !   ##         ##A  !
  !..................!            !   #               !
  !..................!            !   #               !
  !..................!            !   ###             !
  !..................!            !A###     #########A !
  !..................!            !  ##      #        !
  !..................!            !  #        #       !
  !..................!            !  #         #      !
  !..................!            !  #         ##     !
  !..................!            !  #          # ##  !
  !..................!            !A#########    ###A !
  !..................!            !  #               !
  !..................!            !  #         A      !
  --------------------            --------------------

    2000 ITERATIONS                     8000 ITERATIONS
  --------------------            --------------------
  ! A    ............!            ! A        A        !
  ! #    ............!            ! #        #        !
  ! #    ............!            ! #        #        !
  !  # #A............!            !  #  ############  !
  !  ##  ............!            !   ##         ##A  !
  !   #  ............!            !   #               !
  ! .   # ...........!            !   #               !
  !..   ## ..........!            !   ###             !
  !...    # .........!            !A###     #########A !
  !....    A ........!            !  ##      #        !
  !........  ........!            !  #        #       !
  !..................!            !  #        #       !
  !..................!            !  #         #      !
  !..................!            !  #         ##     !
  !..................!            !  #          # ##  !
  !..................!            !A#########    ###A !
  !..................!            !  #               !
                                  !  #         A      !
  --------------------            --------------------

    4000 ITERATIONS                    10000 ITERATIONS
  --------------------            --------------------
  ! A      A        !             ! A        A        !
  ! #      #        !             ! #        #        !
  ! #      #        !             ! #        #        !
  !  # ###########  !             !  #  ###########   !
  !   ##        ##A !             !   ##         ##A  !
  !    #            !             !   #               !
  !    #            !             !   #               !
  !    ###          !             !   ###             !
  !    ##   #######A !            !A###     #########A !
  !    ##    #      !             !  ##      #        !
  !    #      #     !             !  #        #       !
  !    #       #    !             !  #        #       !
  !    #        #   !             !  #         #      !
  !    #        ##  !             !  #         ##     !
  !    ##A  ..  ###A !            !A#########    ###A !
  !      ....       !             !  #               !
  !      ......     !             !  #         A      !
  --------------------            --------------------
```

Simulation 17.4. Formation of net-like structures (see Figs 15.2 and 15.5).

```
0001          SUBROUTINE CONST
0002          COMMON/D/ IC,IPR,KX,KY,KZ,IA,IB,IZ,DIA,DIB,TA,TB,QA,
              1QAA,QB,QBB,QC,QD,QE,QF
0003          DIMENSION ID(8),D(20),DT(20)
0004          EQUIVALENCE (ID(1),IC),(D(1),DIA)
0005          DATA DT/'IC','IPR','KX','KY','KZ','IA','IB','IZ','
              1DIA','DIB','TA','TB','QA','QAA','QB','QBB','QC','QD','QE'
              2,'QF'/
0006        5 WRITE(5,905)
0007          READ(5,908) IS
0008       10 IF (IS.EQ.0) GOTO 600        ! RETURN TO MAIN PROGRAM
0010          IF (IS.LE.8) GOTO 80         ! INPUT INTEGER CONSTANT 1-8
0012          IF (IS.LE.20) GOTO 100       ! INPUT FLOTING POINT CONST. 9-20
0014          IF (IS.EQ.22) CALL EXIT      ! PROGRAM TERMINATION
0016          GOTO 600
0017       80 WRITE (5,909) DT(IS),ID(IS)
0018          READ(5,900)ID(IS)            ! INPUT INTEGER CONSTANT
0019          GOTO 5
0020      100 WRITE (5,910) DT(IS),D(IS-8)
0021          READ (5,902) D(IS-8)         !INPUT FLOATING POINT CONST.
0022          GOTO 5
0023      600 WRITE (5,906) (ID(IK),DT(IK),IK=1,8)
0024          WRITE (5,907) (D(IK-8),DT(IK),IK=9,20)
0025          RETURN
0026      900 FORMAT (I6)
0027      902 FORMAT (F10.4)
0028      905 FORMAT (' NEW CONSTANT # ',$)
0029      906 FORMAT (2(I8,'=',A3),6(I3,'=',A3))
0030      907 FORMAT (6(F8.4,'=',A3),/,6(F8.4,'=',A3))
0031      908 FORMAT (I6,$)
0032      909 FORMAT ('+',A3,' OLD =',I8,'  NEW =?  ',$)
0033      910 FORMAT ('+',A3,' OLD =',F8.4,'  NEW =?  ',$)
0034          END
```

Example:

```
   6000=IC      2000=IPR 18=KX  18=KY    0=KZ   0=IA    0=IB    0=IZ
  0.0200=DIA  0.2000=DIB  0.1200=TA    0.0400=TB    0.0400=QA    0.0040=QAA
  0.0014=QB   0.0200=QBB  0.0300=QC    0.0003=QD    0.0600=QE    0.2000=QF
? 0=NEW CONSTANTS,1=START, 2=CONTINUE, 3=PRINT, 4=END
  0
NEW CONSTANT # 9
DIA OLD =  0.0200   NEW =?  .01                  ___= Input from the
NEW CONSTANT # 12                                       terminal
TB  OLD =  0.0400   NEW =?  .06
NEW CONSTANT # 0
   6000=IC      2000=IPR 18=KX  18=KY    0=KZ   0=IA    0=IB    0=IZ
►0.0100=DIA  0.2000=DIB  0.1200=TA  ►0.0600=TB    0.0400=QA    0.0040=QAA
  0.0014=QB   0.0200=QBB  0.0300=QC    0.0003=QD    0.0600=QE    0.2000=QF
? 0=NEW CONSTANTS,1=START, 2=CONTINUE, 3=PRINT, 4=END
```

Subroutine 17.5. CONST. Allows to change constants individually. Input: constant No. (see Table 17.1) and the new constant. In the example, const. No. 9 and 12 have been changed, see arrows. Constant No. 0 leads to a printout of all 20 constants and to a return into the main program.

In a batchwise computing, the line CALL CONST may be substituted by the following lines (subroutine CONST is then not required).

```
      READ(IFIM,931) IC,IPRINT,KX,KY,KZ,IA,IB,IZ,DIA,DIB,TA,TB,QA,
    1 QAA,QB,QBB,QC,QD,QE,QF
      WRITE (5,931) IC,IPRINT,KX,KY,KZ,IA,IB,IZ,DIA,DIB,TA,TB,QA,
    1 QAA,QB,QBB,QC,QD,QE,QF
  931 FORMAT(1X,2I6,6I3,/,1X,6F8.4,/,1X,6F8.4)
```

```
0001        SUBROUTINE PLOPP (A,AM,KA,KS,IQ)
       C    PRINTS THE CONTENT OF A ONE-DIMENSIONAL ARRAY
       C    IN A SINGLE LINE ON A LINE-PRINTER
       C     IN THE PLOTT, SYMBOLS < .,:;+%$*#> ARE USED FOR 0-10%,10-20%...
       C    RELATIV VALUES, A=ABSOLUTE MAXIMUM
0002        DIMENSION A(1),TY(16),AP(31),M(15)
0003        DATA TY/' ',',','.',';',':',';',';','+','%','$','*','#','A','B',
           1'C','Y','X','N'/
0004        APL=TY(IQ+10)
0005        AM=-100000.
0006        DO 100 I=KA,KS          !   DETERMINATION OF THE MAXIMUM
0007        IF (A(I).LT.AM) GOTO 100
0009        AM=A(I)
0010    100 CONTINUE
0011        AMX=10./(AM+.0000001)
0012        AMXX=100./(AM+.00000001)
0013        IF (KS-KA+1.GT.15) GOTO 305
0015        DO 110 IX=1,15
0016    110 AP(IX)=TY(1)
0017    305 DO 320 IX=KA,KS
0018        N=A(IX)*AMX+1.0001
0019        IF (N.LE.0) N=16
0021        IF(KS.LE.15) M(IX)=A(IX)*AMXX+.5
0023    320 AP(IX)=TY(N)
0024        IF (KS.GT.15)   GOTO 310
0026    210 WRITE(5, 987) AM,APL,(AP(IJ),IJ=1,15),(M(IK),IK=KA,KS)
0027        RETURN
0028    310 WRITE(5, 986) AM,APL,(AP(IX),IX=1,KS)
0029        RETURN
0030    986 FORMAT (1X,F8.4,A1,1X,60A1)
0031    987 FORMAT (1X,F8.4,A1,1X,15A1,1X,15I3)
0032        END
```

Subroutine 17.6. PLOPP. The content of a one-dimensional array is plotted in one line with the line printer. The first number is the maximum value, followed by a letter indicating the plot (e.g. A = activator). The blackness of the symbols indicates the relative concentration. The integer numbers are the relative values in % of the maximum (see Simulation 17.1).

```
0001            SUBROUTINE XYPRINT (A,AM,KAX,KX,KAY,KY,IQ,ITOT)
        C-----PROGRAM FOR XY-PLOT OF A TWO-DIMENSIONAL ARRAY
        C-----WITH A LINE PRINTER
        C     IN THE PLOTT, SYMBOLS < .,:;+%$*#> ARE USED FOR 0-10%,10-20%....
        C     RELATIV VALUES, A=ABSOLUTE MAXIMUM
        C     A:   ARRAY TO BE PLOTTET
        C     AM:  CONTAINS AFTER A CALL THE MAXIMUM OF A
        C     KAX,KX: FIRST AND LAST ELEMENT OF THE X AXIS
        C     KAY,KY: FIRST AND LAST ELEMENT OF THE Y AXIS
        C     IQ: IDTENTIFICATION OF THE PLOTT
        C         E.Q. IQ=11 = PLOT IS FRAMED BY LETTER A
        C     ITOT:TRANSFERED TO THE PLOTT, TOTAL NUMBER OF ITERATION
        C
0002            DIMENSION A(31,31),TY(20),M(31),AP(34)
0003            DATA TY/' ','.',',',';',':',';','+','%','$','*','#','A','B',
               1'S','Y','X','N','1','2','3','4'/
0004          9 AM=-1000.
0005            DO 10 IX=KAX,KX
0006            DO 10 IY=KAY,KY
0007            IF (A(IX,IY).GT.AM)   AM=A(IX,IY)
0009         10 CONTINUE
0010         12 WRITE(5, 900) ITOT,AM
0011            IF (AM.LT.0.000001) RETURN
0013            AMX=10./AM
0014            NX=KX-KAX+3
0015            DO 210 IX=1,34
0016        210 AP(IX)=TY(1)
0017            DO 212 IX=1,NX
0018        212 AP(IX)=TY(IQ)
0019            IF (KX-KAX.LT.16) GOTO 215
0021            WRITE(5, 985)(AP(IX),IX=1,NX)     !UPPER MARGIN OF THE PLOTT
0022            GOTO 220
0023        215 WRITE(5, 945)(AP(IJ),IJ=1,22),((IK),IK=KAX,KX)
0024        220 DO 226 IY=KAY,KY
0025            IXX=2
0026            DO 224 IX=KAX,KX
0027            N=A(IX,IY)*AMX+1.0001     !  N=1...11,DEPENDING ON RELATIVE VALUE
0028            IF (N.LE.0) N=16          !  SELECTS PLOTT SYMBOL
0030            M(IX)=A(IX,IY) *100./AM+.5 ! RELATIVE VALUES IN %
0031            AP(IXX)=TY(N)
0032        224 IXX=IXX+1
0033            AP(1)=TY(IQ)
0034            AP(NX)=TY(IQ)
0035            IF (KX-KAX.LT.16) GOTO 225
0037            WRITE(5, 985)(AP(IX),IX=1,NX)     ! LOWER MARGIN OF THE PLOTT
0038            GOTO 226
0039        225 WRITE(5, 945)(AP(IJ),IJ=1,22),(M(IK),IK=KAX,KX)
0040        226 CONTINUE
0041            DO 230 IX=1,NX
0042        230 AP(IX)=TY(IQ)
0043            WRITE(5, 985)(AP(IX),IX=1,NX)
0044            RETURN
0045        900 FORMAT(X,I5,' ITERATIONS; MAXIMUM= ',F6.2)
0046        945 FORMAT (1X,22A1,16I3)
0047        985 FORMAT (1X,34A1)
0048            END
```

Subroutine 17.7. XYPRINT. Produces a plot of a two-dimensional array on a line printer. Each symbol represents one cell, the blackness indicates relative concentration, see Simulations 17.2 and 17.3.

References

Adler, I. (1974). A model of contact pressure in phyllotaxis. *J. theor. Biol.* **45**, 1–79.

Algire, G. H. and Chalkley, H. W. (1945). Vascular reaction of normal and malignant tissue *in vivo*. I. Vascular reactions of mice to wounds and to normal and neoplastic implants. *J. nat. Cancer Inst.* **6**, 73.

Ashburner, M. and Wright, T. R. F. (eds) (1978). "The Genetics and Biology of *Drosophila*". Academic Press, New York and London.

Avery, G. S. (1933). Structure and development of the tobacco leaf. *Amer. J. Bot.* **20**, 565–592.

Babloyanz, A. and Hiernaux, J. (1975). Models for cell differentiation and generation of polarity in diffusion governed morphogenetic fields. *Bull. Math. Biol.* **37**, 637–657.

Bard, J. B. (1977). A unity underlying the different zebra striping pattern. *J. Zool.* **183**, 527–539.

Bard, J. B. and Lander, I. (1974). How well does Turings theory of morphogenesis work. *J. theor. Biol.* **45**, 501–531.

Bart, A. (1971a). Morphogenese surnumeraire au niveau de la patte du phasme *Carausius morosus* Br. *Wilhelm Roux Arch.* **166**, 331–364.

Bart, A. (1971b). Modalites de formation et de development d'un centre morphogenetique chez *Carausius morosus* Br. *Wilhelm Roux Arch.* **168**, 97–124.

Barth, L. G. (1940). The process of regeneration in hydroids. *Biol. Rev.* **15**, 405–420.

Bateson, W. (1880). On the nature of supernumerary appendages in insects. Proceedings of the Philosophical Society Cambridge, VII. Reprinted in "Scientific Papers of William Bateson" (R. C. Punned, ed.) Vol. I, p. 125. Cambridge University Press, Cambridge (1928).

Berking, S. (1977). Bud formation in Hydra—Inhibition by an endogenous morphogen. *Wilhelm Roux Arch.* **181**, 215–225.

Berking, S. (1979a). Analysis of head and foot formation in hydra by means of an endogenous inhibitor. *Wilhelm Roux Arch.* **186**, 189–210.

Berking, S. (1979b). Control of nerve cell formation from multipotent stem cells in hydra. *J. Cell Sci.* **40**, 193–205.

Bode, H. R. and David, C. N. (1978). Regulation of a multipotent stem cell, the interstitial cell of hydra. *Prog. Biophys. Molec. Biol.* **33**, 189–206.

Bode, P. M. and Bode, H. (1980). Formation of pattern in regenerating tissue pieces of hydra attenuata. I. Head-Body proportion regulation. *Dev. Biol.* **78**, 484–496.

Bode, P. M. and Bode, H. R. (1982). Proportioning a hydra. *Am. Zool.* **22**, 7–15.

Bohn, H. (1965). Analyse der Regenerationsfähigkeit der Insektenextremität durch Amputations- und Transplantationsversuche an Larven der afrikanischen Schabe *Leucophaea maderae* Fabr. (Blattaria). II. Mitt. Achsendetermination. *Wilhelm Roux Arch. Org.* **156**, 449–503.

Bohn, H. (1970a). Interkalare Regeneration und segmentale Gradienten bei den Extremitäten von *Leucophaea*-Larven (Blattaria). I. Femur und Tibia. *Wilhelm Roux Arch.* **165**, 303–341.

Bohn, H. (1970b). Interkalare Regeneration und segmentale Gradienten bei den Extremitäten von *Leucophaea*-Larven (Blattaria). II. Coca und Tarsus. *Dev. Biol.* **23**, 355–379.

Bohn, H. (1971). Interkalare Regeneration und segmentale Gradienten bei den Extremitäten von *Leucophaea*-Larven (Blattaria). III. Die Herkunft des interkalaren Regenerats. *Wilhelm Roux Arch.* **167**, 209–221.

Bohn, H. (1972). The origin of the epidermis in the supernumerary regenerates of triple legs in cockroaches (Blattaria). *J. Embryol. exp. Morphol.* **28**, 185–208.

Bonner, J. T. (1959). Evidence for the sorting out of cells in the development of the cellular slime molds. *Proc. Natl. Acad. Sci. USA* **45**, 379–384.

Bonner, J. T. and Slifkin, M. K. (1949). The proportion of the stalk and spore cells in the slime mold *Dictyostelium discoideum. Amer. J. Bot.* **36**, 727, 1–734.

Boveri, T. (1901). Über die Polarität des Seeigels. *Verh. dt. phys. med. Ges. (Würzburg)* **34**, 145–175.

Bryant, P. J. (1975a). Regeneration and duplication in imaginal discs. *In* "Cell Patterning". Ciba Foundation Symp. No. 29, pp. 71–93. Associated Scientific Publishers, Amsterdam.

Bryant, P. J. (1975b). Pattern formation in the imaginal wing disc of *Drosophila melanogaster*: Fate map, regeneration and duplication. *J. exp. Zool.* **193**, 49–77.

Bryant, P. J. (1978). Pattern formation in the imaginal disc. *In* "The Genetics and Biology of *Drosophila*" (M. Ashburner and T. R. F. Wright, eds) Vol. 2c, pp. 229–335. Academic Press, London and New York.

Bryant, P. J. and Girton, J. R. (1980). Genetics of pattern formation. *In* "Development and Neurobiology of Drosophila" (O. Siddiqui, P. Babu, L. M. Hall and J. C. Hall, eds). Plenum Press, New York and London.

Bryant, S. V. and Baca, B. A. (1978). The regulative ability of double-half and half upper arms in the newt *Notophthalmus viridescens. J. exp. Zool.* **204**, 307–324.

Bryant, S. V. and Iten, L. E. (1976). Supernumerary limbs in amphibians: experimental production in *Notophthalmus viridescens* and a new interpretation of their formation. *Dev. Biol.* **50**, 212–234.

Bryant, S. V., French, V. and Bryant, P. J. (1981). Distal regeneration and symmetry. *Science* **212**, 993–1002.

Bull, A. L. (1966). Bicaudal, a genetic factor which affects the polarity of the embryo in *Drosophila melanogaster. J. exp. Zool.* **161**, 221–242.

Bullière, D. (1972). Etude de la regeneration d'appendice chez un insecte: standes de la formation des regenerats et rapports avec le cycle de mue. *Ann. Embr. Morph.* **5**, 61–74.

Bünning, E., Sagromsky, H. (1948). Die Bildung des Spaltöffnungsmusters in der Blattepidermis. *Z. Naturforsch.* **3b**, 203–216.

Campell, R. D. (1976). Elimination of hydra interstitial and nerve cells by means of colchicine, *J. Cell Sci.* **21**, 1–13.

Chandebois, R. (1973). General mechanisms of regeneration as elucidated by experiments on planarians and by a new formulation of the morphogenetic field concept. *Acta biotheor. (Leiden)* **22**, 2–33.

Chandebois, R. (1976a). "Histogenesis and Morphogenesis in Planarian Regeneration". Monographs in Developmental Biology. Vol. XI. Karger, Basle.

Chandebois, R. (1976b). Cell sociology: a way of reconsidering the current concepts of morphogenesis. *Acta Biotheor. (Leiden)* **25**, 71–102.

Chandebois, R. (1979). The dynamics of wound closure and its role in the programming of planarian regeneration. *Develop. Growth and Differ.* **21**, 195–204.

Child, C. M. (1929). The physiological gradients. *Protoplasma* **5**, 447–476.

Child, C. M. (1941). "Patterns and Problems of Development." Univ. of Chicago Press, Chicago.

Child, C. M. (1946). Organizers in the development and the organizer concept. *Physiol. Zoology* **19**, 89–148.

Chung, S. H. and Cooke, J. (1975). Polarity of structure and of ordered nerve connections in the developing amphibian brain. *Nature* **258**, 126–132.

Cohen, M. H. and Robertson, A. (1971). Wave propagation in early stages of aggregation of cellular slime molds. *J. theor. Biol.* **31**, 101–118.

Cooke, J. (1975a). Control of somite number during morphogenesis of a vertebrate, *Xenopus laevis. Nature* **254,** 196–199.

Cooke, J. (1975b). The emergence and regulation of spatial organization in early animal development. *Ann. Rev. Biophys. Bioeng.* **4**, 185–217.

Cooke, J. (1981a). Scale of body pattern adjusts to available cell number in amphibian embryos. *Nature* **290**, 775–777.

Cooke, J. (1981b). The problem of periodic patterns in embryos. *Phil. Trans. R. Soc. Lond.* B **295**, 509–524.

Cooke, J. and Zeemann, E. C. (1976). A clock and wavefront model for control of the number of repeated structures during animal morphogenesis. *J. theor. Biol.* **58**, 455–476.

Counce, S. (1973). The causal analysis of insect embryogenesis. *In* "Developmental Systems: Insects." (S. Counce and H. C. Waddington, eds). Vol. II, pp. 1–156. Academic Press, London and New York.

Counce, S. J. and Waddington, C. H. (eds) (1972). "Developmental Systems: Insects." Academic Press, London and New York.

Crick, F. (1970). Diffusion in embryogenesis. *Nature* **225**, 420–422.

Crick, F. H. C. and Lawrence, P. A. (1975). Compartments and polyclones in insect development. *Science* **189**, 340–347.

Czihak, G. (ed.) (1975). "The Sea Urchin Embryo." Springer, Heidelberg.

Deuchar, E. M. and Burges, A. M. C. (1967). Somite segmentation in amphibian embryos: is there a transmitted control mechanism? *J. Embryol. exp. Morph.* **17**, 349–358.

Driesch, H. (1899). Die Lokalisation morphogenetischer Vorgänge. Ein Beweis vitalistischen Geschehens. *Arch. f. Entw. Mech.* **8**, 35–111.

Driesch, H. (1900). Studien über das Regulationsvermögen der Organismen. 4. Die Verschmelzung der Individualität bei Echinidenkeimen. *Arch. f. Entw. Mech.* **10**, 411–434.

Durston, A. J. and Vork, T. (1979). A kinematic study of the development of vitally stained *Dictyostelium discoideum. J. Cell Sci.* **36**, 261–279.

Ehrenstein, G. v. and Schierenberg, E. (1980). Cell lineages and development of *Caernorhabditis elegans* and other nematodes. *In* "Nematodes as Biological Models" (B. Zuckerman, ed.) Vol. 1, pp. 1–71. Academic Press, New York and London.

Eigen, M. and Schuster, P. (1978). The hypercycle. A principle of self-organization. Part B: The abstract hypercycle. *Naturwissenschaften* **65**, 7–41.

Faber, J. (1976). Positional information in the amphibian limb. *Acta Biotheor. Leiden* **25**, 44–65.

Fallon, J. F. and Crosby, G. M. (1975). Normal development of the chick wing following removal of the polarizing zone. *J. exp. Zool.* **193**, 449–455.

Fife, P. C. (1979). "Mathematical Aspects of Reacting and Diffusing Systems". Lecture Notes in Biomathematics. Springer, Heidelberg.

Fisher, P. R., Smith, E. and Williams, K. L. (1981). An extracellular chemical signal controlling phototactic behaviors by *D. discoideum* slugs. *Cell* **23**, 799–807.

Folkman, J. (1976). The vascularization of a tumor. *Sci. American* **234**, May, 58–73.

Folkman, J., Merler, E., Abernathy, C. and Williams, G. (1971). Isolation of a tumor factor responsible for angiogenesis. *J. exp. Med.* **133**, 275–288.

Fraser, S. E. (1980). A differential adhesion approach to the patterning of nerve connection. *Dev. Biol.* **79**, 453–464.

French, V. (1976a). Leg regeneration in the cockroach, *Blattella germanica* I. Regeneration from a congruent tibial graft/host junction. *Wilhelm Roux Arch.* **179**, 57–76.

French, V. (1976b). Leg regeneration in the cockroach, *Blattella germanica*. II. Regeneration from non-congruent graft/host junction. *J. Embryol. exp. Morph.* **35**, 267–301.

French, V. (1978). Intercalary regeneration around the circumference of the cockroach leg. *J. Embryol. exp. Morph.* **47**, 53–84.

French, V. (1980). Positional information around the segments of the cockroach leg. *J. Embryol. exp. Morph.* **59**, 281–313.

French, V., Bryant, P. J. and Bryant, S. V. (1976). Pattern regulation in epimorphic fields. *Science* **193**, 969–981.

Garcia-Bellido, A. (1975). Genetic control of wing disc development in *Drosophila*. *In* "Cell Patterning." Ciba Foundation Symp. No. 29, pp. 161–182. Associated Scientific Publishers, Amsterdam.

Garcia-Bellido, A. (1977). Homoeotic and atavic mutations in insects. *Amer. Zool.* **17**, 613–629.

Garcia-Bellido, A. and Nöthiger, R. (1976). Maintenance of determination by cells of imaginal discs of *Drosophila* after dissociation and culture in vivo. *Wilhelm Roux Arch.* **180**, 189–206.

Garcia-Bellido, A., Ripoll, P. and Morata, G. (1973). Developmental compartmentalization of the wing disc of *Drosophila*. *Nature new Biol.* **245**, 251–253.

Garcia-Bellido, A., Ripoll, P. and Morata, G. (1976). Developmental compartmentalization in the dorsal mesothoracic disc of *Drosophila*. *Dev. Biol.* **48**, 132–147.

Gasseling, M. T. and Saunders, J. W., Jr. (1964). Effect of the "Posterior Necrotic Zone" of the early chick wing bud on the pattern and symmetry of limb outgrowth. *Amer. Zool.* **4**, 303–304.

Gaze, R. M., Feldman, J. D., Cooke, J. and Chung, S. H. (1979). The orientation of the visuotectal map in *Xenopus*: developmental aspects. *J. Embryol. exp. Morph.* **53**, 39–66.

Gehring, W. J. and Nöthiger, R. (1973). The imaginal disc of *Drosophila. In* "Developmental Systems: Insects." (S. J. Counce and H. C. Waddington, eds), Vol. II, pp. 211–290. Academic Press, New York and London.

Gerisch, G. (1968). Cell aggregation and differentiation in *Dictyostelium. Curr. Top. Dev. Biol.* **3**, 157–232.

Gerisch, G. and Hess, B. (1974). Cyclic AMP-controlled oscillation in suspended *Dictyostelium* cells: Their relation to morphogenetic cell interactions. *Proc.Natl. Acad. Sci. (Wash.)* **71**, 2118–2122.

Gierer, A. (1977a). Biological features and physical concepts of pattern formation exemplified by hydra. *Curr. Top. Dev. Biol.* **11**, 17–59.

Gierer, A. (1977b). Physical aspects of tissue evagination and biological form. *Quarterly Rev. Biophys.* **10**, 529–593.

Gierer, A. (1981a). Generation of biological patterns and form: Some physical, mathematical. and logical aspects. *Prog. Biophys. molec. Biol.* **37**, 1–47, 1980.

Gierer, A. (1981b). Development of projections between areas of the nervous system. *Cybernetics* **42**, 69–78.

Gierer, A. (1981c). Socioeconomic inequalities: Effect of selfenhancement, depletion and redistribution. *Jahrb. f. Nationalök. u. Stat.* **196**, 309–331.

Gierer, A. and Meinhardt, H. (1972). A theory of biological pattern formation. *Kybernetik* **12**, 30–39.

Gierer, A. and Meinhardt, H. (1974). Biological pattern formation involving lateral inhibition. "Lectures on Mathematics in the Life Science", No. 7, 163–183. The American Mathematical Society, Providence, Rhode Island.

Gierer, A., Berking, S., Bode, H., David, C. N., Flick, K., Hansmann, G., Schaller, H. and Trenkner, E. (1972). Regeneration of *hydra* from reaggregated cells. *Nature New Biol.* **239**, 98–101.

Girton, J. R. (1981). Pattern triplication produced by a cell-lethal mutation in *Drosophila. Dev. Biol.* **84**, 164–172.

Girton, J. R. and Russell, M. A. (1981). An analysis of compartmentalization in pattern duplications induced by a cell-lethal mutation in *Drosophila. Dev. Biol.* **85**, 55–64.

Gmitro, J. I. and Scriven, L. E. (1966). A physiochemical basis for pattern and rhythm. *In* "Intracellular Transport", p. 221. Academic Press, New York and London.

Goodwin, B. C. and Cohen, H. M. (1969). A phase-shift model for the spatial and temporal organization of developing systems. *J. theor. Biol.* **25**, 49–107.

Graf, L. and Gierer, A. (1980). Size, shape and orientation of cells in budding hydra and regulation of regeneration in cell aggregates. *Wilhelm Roux Arch.* **188**, 141–151.

Granero, M. I., Porati, A. and Zanacca, D. (1977). A bifurcation analysis of pattern formation in a diffusion governed morphogenetic field. *J. Math. Biology* **4**, 21–27.

Gross, J. D., Town, C. D., Brookman, J. J., Jermyn, K. A., Peacey, M. J., and Kay, R. R. (1981). Cell patterning in *Dictyostelium. Phil. Trans. R. Soc. Lond.* B **295**, 497–508.

Hadorn, E. (1967). Dynamics of determination. *In* "Major Problems of Developmental Biology" (M. Locke, ed.), pp. 85–104. Academic Press, London and New York.

Harrison, R. G. (1910). The outgrowth of the nerve fibre as a mode of protoplasmic movements. *J. exp. Zool.* **9**, 787–846.

Harrison, R. G. (1921). On relations of symmetry in transplanted limbs. *J. exp. Zool.* **32**, 1–136.

Hausen, K., Wolburg-Buchholz, K., and Ribi, W. A. (1980). The synaptic organization of visual interneurons in the lobula complex of flies. *Cell Tissue Res.* **208**, 371–387.

Hayes, P. H., Girton, J. R. and Russell, M. A. (1979). Positional information and the bithorax-complex. *J. theor. Biol.* **79**, 1–17.

Haynie, J. L. and Bryant, P. J. (1976). Intercalary regeneration in imaginal wing disc of *Drosophila melanogaster*. *Nature* **259**, 659–662.

Haynie, J. and Schubiger, G. (1979). Absense of distal to proximal intercalary regeneration in the imaginal wing discs of *Drosophila melanogaster*. *Dev. Biol.* **68**, 151–161.

Herbst, C. (1942). Hans Driesch als experimenteller und theoretischer Biologe. *Wilhelm Roux Arch.* **141**, 111–153.

Hertel, R. and Flory, R. (1968). Auxin movement in corn coleoptiles. *Planta* **82**, 123–144.

Herth, W. and Sander, K. (1973). Mode and timing of body pattern formation in the early embryonic development of cyclorrhaphic dipterans (*Protophormia, Drosophila*). *Wilhelm Roux Arch.* **172**, 1–27.

Hicklin, J., Hornbruch, A., Wolpert, L., and Clarke, M. (1973). Positional information and pattern regulation in hydra: the formation of boundary regions following axial grafts. *J. Embryol. exp. Morph.* **30**, 701–725.

Hinchliffe, J. R. and Johnson, D. R. (1980). "The Development of the Vertebrate Limb." Oxford University Press, London.

Holder, N., Tank, P. W. and Bryant, S. V. (1980). Regeneration of symmetrical forelimbs in the axolotl *Ambystoma mexicanum*. *Dev. Biol.* **74**, 302–314.

Hörstadius, S. (1973). "Experimental Embryology of Echinoderms." Clarendon Press, Oxford.

Hubel, D. H., Wiesel, T. N. and LeVay, S. (1977). Plasticity of ocular dominance columns in monkey striate cortex. *Phil. Trans. Roy. Soc. London B* **278**, 377–409.

Iten, L. E., and Bryant, S. V. (1975). The interaction between the blastema and stump in the establishment of the anterior-posterior and proximal-distal organization of the limb regenerate. *Dev. Biol.* **44**, 119–147.

Iterson, G. van. (1907). "Mathematische und mikroskopisch-anatomische Studien über Blattstellungen," Fischer, Jena.

Jaffe, F. (1968). Localization in the developing *Fucus* egg and the general role of localizing currents. *Adv. Morphogenesis* **7**, 295–328.

Jost, L. (1942). Über Gefässbrücken. *Z. Bot.* **38**, 161–215.

Jürgens, G. and Gateff, E. (1979). Pattern specification in imaginal discs of *Drosophila melanogaster*. Developmental analysis of a temperature-sensitive mutant producing duplicated legs. *Wilhelm Roux Arch.* **186**, 1–25.

Jung, E. (1966). Untersuchungen am Ei des Speisebohnen-Käfers *Bruchiduis obtectus* SAY (Coleoptera) II. Entwicklungsphysiologische Ergebnisse der Schnürungs-experimente. *Wilhelm Roux Arch.* **157**, 320–392.

Kalthoff, K. (1971). Revisionsmöglichkeiten der Entwicklung zur Missbildung "Doppelabdomen" im partiell UV-bestrahlten Ei von *Smittia* spec. (Diptera, Chironomidae) *Zool. Anz. Suppl.* **34**, 61–65.

Kalthoff, K. (1976). Specification of the antero-posterior body pattern in insect eggs. *In* "Insect Development" (P. A. Lawrence, ed.) pp. 53–75. Blackwell Scientific Publ., Oxford.

Kalthoff, K. and Sander, K. (1968). Der Entwicklungsgang der Missbildung "Doppelabdomen" im partiell UV-bestrahlten Ei von *Smittia parthenogenetica* (Diptera, Chironomidae). *Wilhelm Roux Arch.* **161**, 129–146.

Karlsson, J. (1980). Distal regeneration in proximal fragments of the wing disc of *Drosophila. J. Embryol. exp. Morph.* **59**, 315–323.

Kauffman, S. A. and Ling, E. (1981). Regeneration by complementary wing disc fragments of *Drosophila melanogaster. Dev. Biol.* **82**, 238–257.

Kauffman, S. A., Shymko, R. M. and Trabert, K. (1978). Control of sequential compartment formation in *Drosophila. Science* **199**, 259–270.

Kieny, M. (1960). Role inducteur du mesoderme dans la differentiation precoce du bourgeon de membre chez l'embryon de poulet. *J. Embryol. exp. Morph.* **8**, 457–467.

Kimble, J. E. (1981). Strategies for control of pattern formation in *Caenorhabditis elegans. Phil. Trans. R. Soc. Lond. B* **295**, 539–551.

Kochav, Sh. and Eyal-Giladi, H. (1971). Bilateral symmetry in chick embryo determination by gravity. *Science* **171**, 1027–1029.

Kopell, N. J. and Howard, L. N. (1973). Plane wave solutions to reaction-diffusion equations. *Studies in Appl. Math.* **52**, 291–328.

Krasner, G. N. and Bryant, S. V. (1980). Distal transformation from double-half forearms in the axolotl, *Ambystoma mexicanum. Dev. Biol.* **74**, 315–325.

Krause, G. (1939). Die Eitypen der Insekten. *Biol. Zbl.* **59**, 495–536.

Kühn, A. (1948). Die Wirkung der Mutation Va bei *Ptychopoda seriata Z. indukt. Abstamm.- u. Vererb.-Lehre* **82**, 430–447.

Lacalli, T. C. and Harrison, L. G. (1979). Turings condition and the analysis of morphogenetic models. *J. theor. Biol.* **76**, 419–436.

Landström, U. (1977). On the differentiation of prospective ectoderm to a ciliated cell pattern in embryos of *Ambystoma mexicanum. J. Embryol. exp. Morph.* **41**, 23–32.

Lawrence, P. A. (1966a). Gradients in the insect segment: the orientation of hairs in the weed bug *Oncopeltus fasciatus. J. exp. Biol.* **49**, 607–620.

Lawrence, P. A. (1966b). Development and determination of hairs and bristles in the milkweed bug *Oncopeltus fasciatus. J. Cell. Sci.* **1**, 475–498.

Lawrence, P. A. (1970). Polarity and patterns in the postembryonic development of insects. *Adv. Insect Physiol.* **7**, 197–266.

Lawrence, P. A. (1973). A clonal analysis of segment development in *Oncopeltus* (Hemiptera). *J. Embryol. exp. Morph.* **30**, 681–699.

Lawrence, P. A., Green, S. M. and Johnston, P. (1978). Compartmentalization and growth of the *Drosophila* abdomen. J. Embryol. exp. Morph. **43**, 233–245.

Lawrence, P. A., Struhl, G. and Morata, G. (1979). Bristle patterns and compartment boundary in the tarsi of *Drosophila. J. Embryol. exp. Morph.* **51**, 195–208.

Leach, C. K., Ashworth, J. M. and Garrod, D. R. (1973). Cell sorting out during the differentiation of mixtures of metabolically distinct populations of *Dictyostelium discoideum. J. Embryol. exp. Morphol.* **29**, 647–661.

Levi-Montalcini, R. (1964). Growth control of nerve cells by a protein factor and its antiserum. *Science* **143**, 105–110.

Lewis, E. B. (1963). Genes and developmental pathways. *Amer. Zool.* **3**, 33–56.

Lewis, E. B. (1964). Genetic control and regulation of developmental pathways. *In* "Role of Chromosomes in Development" (M. Locke, ed.) pp. 231–252. Academic Press, New York and London.

Lewis, E. B. (1978). A gene complex controlling segmentation in *Drosophila. Nature* **276**, 565–570.

Lewis, J., Slack, J. and Wolpert, L. (1977). Thresholds in development. *J. theor. Biol.* **65**, 579–590.

Lindahl, P. E. (1932). Zur experimentellen Analyse der Determination der Dorso-ventralachse beim Seeigelkeim. *Wilhelm Roux Arch.* **127**, 300–321.

Locke, M. (1959). The cuticular pattern in an insect, *Rhodnius prolixus* Stal. *J. exp. Biol.* **36**, 459–477.

Lohs-Schardin, M. and Sander, K. (1976). A dicephalic monster embryo of *Drosophila melanogaster*. *Wilhelm Roux Arch.* **179**, 159–162.

Lohs-Schardin, M., Cremer, C. and Nüsslein-Volhard, C. (1979). A fate map for the larval epidermis of *Drosophila melanogaster*: Localized cuticle defects following irradiation of the blastoderm with an ultraviolet laser microbeam. *Dev. Biol.* **73**, 239–255.

Loomis, W. F. (1975). Dictyostelium Discoideum—A Developmental System. Academic Press, New York and London.

Macagno, E. R., Lopresti, V. and Levinthal, C. (1973). Structure and development of neuronal connections in isogenetic organisms: Variations and similarities in the optic system of *Daphnia magua*. *Proc. Natl Acad. Sci.* **70**, 57–61.

MacCabe, J. A., Saunders, J. W., Jr. and Pickett, M. (1973). The control of anteroposterior and dorsoventral axis in embryonic chick limb constructed of dissociated and reaggregated limb-bud mesoderm. *Dev. Biol.* **31**, 323–335.

MacCabe, J. A., Errick, J. and Saunders, J. W., Jr. (1974). Ectodermal control of the dorsoventral axis of the leg bud of the chick embryo. *Dev. Biol.* **39**, 69–82.

MacWilliams, H. K. and Bonner, J. T. (1979). The prestalk–prespore pattern in cellular slime molds. *Differentiation* **14**, 1–22.

MacWilliams, H. K., Kafatos, F. C. and Bossert, W. H. (1970). The feedback inhibition of basal disc regeneration in hydra has a continuously variable intensity. *Dev. Biol.* **23**, 380–398.

Maden, M. (1980). Structure of supernumerary limbs. *Nature* **286**, 803–805.

Maden, M. (1981). Experiments on anuran limb buds and their significance for principles of vertebrate limb development. *J. Embryol. exp. Morph.* **63**, 243–265.

Maden, M. (1982). Supernumerary limbs in amphibians. *Am. Zool.* **22**, 131–142.

Maeda, Y. and Maeda, M. (1974). Heterogeneity of the cell population of the cellular slime mold *Dictyostelium discoideum* before aggregation and its relation to subsequent locations of the cells. *Exp. Cell Res.* **84**, 88–94.

Malchow, D., Nanjundiah, V., Wurster, B., Eckstein, F. and Gerisch, G. (1978). Cyclic AMP-induced pH change in *Dictyostelium discoideum* and their control by calcium. *Biochim. Biophys. Acta* **538**, 473–480.

Martinez, H. M. (1972). Morphogenesis and chemical dissipative structures, a computer simulated case study. *J. theor. Biol.* **36**, 479–501.

Maruyama, M. (1963). The second cybernetics deviation-amplifying mutual causal processes. *American Scientist* **51**, 164–179.

Matsukuma, S., Durston, A. J. (1979). Chemotactic cell sorting in *Dictyostelium discoideum*. *J. Embryol. exp. Morph.* **50**, 243–251.

Meinhardt, H. (1974). The formation of morphogenetic gradients and fields. *Ber. dtsch. bot. Ges.* **87**, 101–108.

Meinhardt, H. (1976). Morphogenesis of lines and nets. *Differentiation* **6**, 117–123.

Meinhardt, H. (1977). A model for pattern formation in insect embryogenesis. *J. Cell Sci.* **23**, 117–139.

Meinhardt, H. (1978a). Models for the ontogenetic development of higher organisms. *Rev. Physiol. Biochem. Pharmacol.* **80**, 48–104.

Meinhardt, H. (1978b). Space-dependent cell determination under the control of a morphogen gradient. *J. theor. Biol.* **74**, 307–321.

Meinhardt, H. (1980). Cooperation of compartments for the generation of positional information. *Naturforsch.* **35c**, 1086–1091.

Meinhardt, H. (1981). The role of compartmentalization in the activation of particular control genes and in the generation of proximo-distal positional information. *Am. Zool.*, in press.

Meinhardt, H. and Gierer, A. (1974). Application of a theory of biological pattern formation based on lateral inhibition. *J. Cell Sci.* **15**, 321–346.

Meinhardt, H. and Gierer, A. (1980). Generation and regeneration of sequences of structures during morphogenesis. *J. theor. Biol.* **85**, 429–450.

Merker, H. J., Nau, H. and Neubert, D. (eds) (1980). "Teratology of the Limbs." Walter de Gruyter, Berlin and New York.

Mitchison, G. J. (1977). Phyllotaxis and the Fibonacci series. *Science* **196**, 270–275.

Mitchison, G. J. (1980). A model for vein formation in higher plants. *Proc. R. Soc. Lond. B* **207**, 79–109.

Morata, G. and Lawrence, P. A. (1977). The development of wingless, a homeotic mutation of *Drosophila*. *Dev. Biol.* **56**, 227–240.

Morgan, T. H. (1901). "Regeneration." Macmillan, New York and London.

Morgan, T. H. (1904). An attempt to analyse the phenomena of polarity in tubularia. *J. exp. Zool.* **1**, 587–591.

Müller, W. A. and Plickert, G. (1982). Quantitative analysis of an inhibitory gradient field in the hydrozoan stolon. (Submitted for press).

Murray, J. D. (1981). A prepattern formation mechanism for animal coat markings. *J. theor. Biol.* **88**, 161–199.

Newman Jr., S. M. and Schubiger, G. (1980). A morphological and developmental study of *Drosophila* embryos ligated during nuclear multiplication. *Dev. Biol.* **79**, 128–138.

Nitschmann, J. (1959). Segmentverluste beim geschnürten *Calliphora*-Keim. *Zool. Anz. Suppl.* **22**, 370–377.

Nübler-Jung, K. (1977). Pattern Stability in the Insect Segment. I. Pattern Reconstitution by Intercalary Regeneration and Cell Sorting in *Dysdercus intermedius* Dist. *Wilhelm Roux Arch* **183**, 17–40.

Nüsslein-Volhard, C. (1977). Genetic analysis of pattern formation in the embryo of *Drosophila melanogaster*. *Wilhelm Roux Arch.* **183**, 249–268.

Nüsslein-Volhard, C. (1979). Maternal effect mutations that after the spatial coordinates of the embryo of *Drosophila melanogaster*. *In* "Determination of spatial organization" (S. Subtelney and I. R. Konigsberg) p. 185–211. Academic Press, New Tork and London.

Nüsslein-Volhard, C. and Wieschaus, E. (1980). Mutants affecting segment number and polarity in *Drosophila*. *Nature* **287**, 795–801.

Nüsslein-Volhard, C., Lohs-Schardin, M., Sander, K. and Cremer, C. (1980). A dorso-ventral shift of embryonic primordia in a new maternal effect mutant of *Drosophila*. *Nature* **283**, 474–476.

Patten, B. M. (1958). "Foundations of Embryology." McGraw-Hill, New York.

Pearson, M. and Elsdale, T. (1979). Somitogenesis in amphibian embryos. *J. Embryol. exp. Morph.* **51**, 27–50.

Pescitelli, M. J. Jr. and Stocum, D. L. (1980). The origin of skeletal structures during intercalary regeneration of larval *Ambystoma* limbs. *Dev. Biol.* **79**, 255–275.

Plickert, G. (1980). Mechanically induced stolon branching in *Eirene viridula* (*Thecata, Campanulinidae*). "Development and Cellular Biology of Coelenterates" (P. Tardent and R. Tardent, eds). Elsevier/North-Holland, Biomedical Press.

Postlethwait, J. H. (1978). Development of cuticular pattern in the legs of a cell lethal mutant in *Drosophila melanogaster*. *Wilhelm Roux Arch.* **185**, 37–57.

Prigogine, I. and Lefever, R. (1968). Symmetry breaking instabilities in dissipative systems. II. *J. chem. Phys.* **48**, 1695–1700.

Prigogine, I. and Nicolis, G. (1971). Biological order, structure, and instabilities. *Quart. Rev. Biophys.* **4**, 107–148.

Ptashne, M., Jeffrey, A., Johnson, A. D., Maurer, R., Meyer, B. J., Pabo, C. O., Roberts, T. M. and Sauer, R. T. (1980). How the Lambda Repressor and Cro work. *Cell* **19**, 1–11.

Raper, K. B. (1940). Pseudoplasmodium formation and organization in *Dictyostelium discoideum*. *J. Elisha Mitchell Sci. Soc.* **56**, 241–282.

Rau, K. G. and Kalthoff, K. (1980). Complete reversal of antero-posterior polarity in a centrifuged insect embryo. *Nature* **287**, 635–637.

Richards, F. J. (1948). The geometry of phyllotaxis and its origin. *Symp. Soc. exp. Biol.* **2**, 217.

Richter, P. H. and Schranner, R. (1978). Leaf Arrangement. Geometry, Morphogenesis, and Classification. *Naturwissenschaften* **65**, 319–327.

Ripley, S. and Kalthoff, K. (1981). Double abdomen induction with low UV-dose in *Smittia* Spec. (Chironomidae, Diptera): Sensitive period and complete photoreversibility. *Wilhelm Roux Arch.* **190**, 49–54.

Robertson, A. and Cohen, A. D. (1971). Control of developing fields. *Ann. Rev. Biophys. Bioeng.* **1**, 409–464.

Rubin, L. and Saunders, J. W. Jr. (1972). Ectodermal-mesodermal interaction in the growth of limb buds in the chick embryo: constancy and temporal limits of the ectodermal induction. *Dev. Biol.* **28**, 95–112.

Runnström, J. (1929). Über Selbstdifferenzierung und Induktion beim Seeigelkeim. *Wilhelm Roux Arch.* **117**, 123–145.

Russel, M. A., Girton, J. R. and Morgan, K. (1977). Pattern formation in a ts-cell-lethal mutant of *Drosophila*: The range of phenotypes induced by larval heat treatment. *Wilhelm Roux Arch.* **183**, 41–59.

Sachs, T. (1975). The control of differentiation of vascular networks. *Ann. Bot.* **39**, 197–204.

Sander, K. (1959). Analyse des ooplasmatischen Reaktions-systems von *Euscelis plebejus* Fall. (Cicadina) durch Isolieren und Kombinieren von Keimteilen. I. Mitt.: Die Differenzierungsleistungen orderer und hinterer Eiteile. *Wilhelm Roux Arch.* **151**, 430–497.

Sander, K. (1960). Analyse des ooplasmatischen Reaktions-systems von *Euscelis plebejus* Fall. (Cicadina) durch Isolieren und Kombinieren von Keimteilen. II. Mitt.: Die Differenzierungsleistungen nach Verlagern von Hinterpolmaterial. *Wilhelm Roux Arch.* **151**, 660–707.

Sander, K. (1961a). New experiments concerning the ooplasmic reaction system of *Euscelis plebejus* (Cicadina). *In* "Symp. on Germ and Development", pp. 338–353. Inst. Int. Embryol. & Fondazione Baselli Pallanza.

Sander, K. (1961b). Umkehr der Keimstreifpolarität in Eifragmenten von *Euscelis* (Cicadina). *Experientia* **17**, 179–180.

Sander, K. (1962). Über den Einfluss von verlagertem Hinterpolmaterial auf das metamere Organisationsmuster im Zikaden-Ei. *Zool. Anz. Suppl.* **25**, 315–322.

Sander, K. (1971). Pattern formation in longitudinal halves of leaf hopper eggs (Homoptera) and some remarks on the definition of "Embryonic regulation". *Wilhelm Roux Arch.* **167**, 336–352.

Sander, K. (1975a). Bildung und Kontrolle räumlicher Muster bei Metazoen. *Verh. dtsch. zool. Ges.* **67**, 58–70.

Sander, K. (1975b). Pattern specification in the insect embryo. *In* "Cell Patterning." Ciba Foundation Symp. No. 29, pp. 241–263. Associated Scientific Publishers, Amsterdam.

Sander, K. (1976). Formation of the basic body pattern in insect embryogenesis. *Adv. Insect Physiol.* **12**, 125–238.

Sander, K. (1981). Pattern generation and pattern conservation in insect ontogenesis—problems, data and models. *Fortschritte der Zoologie* **26**, 101–119.

Sander, K., Lohs-Schardin, M. and Baumann, M. (1980). Embryogenesis in a *Drosophila* mutant expressing half the normal segment number. *Nature* **287**, 841–843.

Saunders, J. W., Jr. (1948). The proximo-distal sequence of the origin of the parts of the chick wing and the role of the ectoderm. *J. exp. Zool.* **108**, 363–403.

Saunders, J. W., Jr. (1969). The interplay of morphogenetic factors. *In* "Limb Development and Deformity" (C. A. Swinyard, ed.), pp. 84–100, Chas. C. Thomas, Illinois.

Saunders, J. W., Jr. (1977). The experimental analysis of chick limb bud development. *In* "Vertebrate Limb and Somite Morphogenesis" (D. A. Ede, J. R. Hinchliffe, and M. Balls, eds) pp. 1–24. Cambridge University Press, Cambridge.

Saxen, L. and Toivonen, S. (1962). Primary Embryonic Induction. Academic Press, London and New York.

Schaller, H. C. (1973). Isolation and characterization of a low-molecular-weight substance activating head and bud formation in hydra. *J. Embryol. exp. Morph.* **29**, 27–38.

Schaller, H. C. (1981). Morphogenetic substances in hydra. *Fortschr. Zool.* **26**, 153–162.

Schaller, H. C. (1982). To be published.

Schaller, H. C. and Gierer, A. (1973). Distribution of the head-activating substance in hydra and its localization in membranous particles in nerve cells. *J. Embryol. exp. Morph.* **29**, 39–52.

Schmidt, O., Zissler, D., Sander, K. and Kalthoff, K. (1975). Switch in pattern formation after puncturing the anterior pole of *Smittia* eggs (Chironomidae, Diptera). *Dev. Biol.* **46**, 216–221.

Schoute, J. C. (1913). Beiträge zur Blattstellung. *Rec. trav. bot. Neerl.* **10**, 153–325.

Schubiger, G. (1968). Anlageplan, Determinationszustand und Transdeterminationsleistungen der männlichen Vorderbeinscheibe von *Drosophila melanogaster*. *Wilhelm Roux Arch.* **160**, 9–40.

Schubiger, G. (1971). Regeneration, duplication and transdetermination in fragments of the leg disc of *Drosophila melanogaster*. *Dev. Biol.* **26**, 277–295.

Schubiger, G. and Wood, W. J. (1977). Determination during early embryogenesis in *Drosophila melanogaster*. *Amer. Zool.* **17**, 565–576.

Schubiger, G. and Schubiger, M. (1978). Distal Transformation in *Drosophila* leg imaginal disc fragments. *Dev. Biol.* **67**, 286–296.

Schwabe, W. W. (1971). Chemical modification of phyllotaxis and its implications. *Symp. Soc. exp. Biol.* **25**, 301–322.

Schwarz, U., Ryter, A., Rambach, A., Hellio, R. and Hirota, Y. (1975). Process of Cellular Division in *Escherichia coli*. *J. mol. Biol.* **98**, 749–759.

Seidel, F. (1929). Untersuchungen über deas Bildungsprinzip der Keimanlage im Ei der Libelle *Platycnemis pennipes*. *Wilhelm Roux Arch.* **119**, 322–440.

Seidel, F. (1935). Der Anlagen-Plan im Libellen-Ei. *Wilhelm Roux Arch.* **132**, 671–751.

Slack, J. M. W. (1976). Determination of polarity in the amphibian limb. *Nature* **261**, 44–46.

Slack, J. M. W. (1977a). Determination of antero-posterior polarity in the axolotl forelimb by an interaction between limb and flank rudiments. *J. Embryol. exp. Morph.* **39**, 151–168.

Slack, J. M. W. (1977b). Control of antero-posterior pattern in the axolotl forelimb by a smoothly graded signal. *J. Embryol. exp. Morph.* **39**, 169–182.

Slack, J. M. W. and Savage, S. (1978a). Regeneration of reduplicated limbs in contravention of the complete circle rule. *Nature* **271**, 760–761.

Slack, J. M. W. and Savage, S. (1978b). Regeneration of mirror symmetrical limbs in axolotl. *Cell* **14**, 1–8.

Snow, M. and Snow, R. (1935). *Philos. Trans. B* **225**, 63.

Sondhi, K. C. (1963). The biological foundations of animal pattern. *Quart. Rev. Biol.* **38**, 289–327.

Spemann, H. (1938). "Embryonic Development and Induction." Yale University Press, New Haven.

Spemann, H. and Mangold, H. (1924). Über Induktion von Embryonalanlagen durch Implantation artfremder Organisatoren. *Wilhelm Roux Arch.* **100**, 599–638.

Steiner, E. (1976). Establishment of compartments in the developing leg discs of *Drosophila melanogaster*. *Wilhelm Roux Arch.* **180**, 9–30.

Stern, C. (1956). Genetic mechanisms in the localized initiation of differentiation, *Symp. Quant. Biol.* **21**, 375–382.

Stocum, D. L. (1968). The urodele limb regeneration blastema: a selforganizing system. II. Morphogenesis and differentiation of autografted whole and fractional blastemas. *Dev. Biol.* **18**, 457–480.

Stocum, D. L. (1975a). Outgrowth and pattern formation during limb ontogenie and regeneration. *Differentiation* **3**, 167–182.

Stocum, D. L. (1975b). Regulation after proximal or distal transposition of limb regeneration blastemas and the determination of the proximal boundary of the regenerate. *Dev. Biol.* **45**, 112–136.

Stocum, D. L. (1978). Regeneration of symmetrical hindlimbs in larval salamanders. *Science* **200**, 790–793.

Strub, S. (1977a). Pattern regulation and transdetermination in *Drosophila* imaginal leg disc reaggregates. *Nature* **296**, 688–691.

Strub, S. (1977b). Development potentials of the cells of the male foreleg disc of *Drosophila*. II. Regulative behaviour of dissociated fragments. *Wilhelm Roux Arch.* **182**, 75–92.

Strub, S. (1979). Heteromorphic regeneration in the developing imaginal primordia of *Drosophila*. *In* "Cell Lineage, Stem Cells and Cell Determination; INSERM Symposium No. 10 (N. Le Douarin, ed.) pp. 311–324, North Holland, Amsterdam.

Sugiyama, T. (1981). Roles of head-activation and head-inhibition potentials in pattern formation of hydra: analysis of a multi-headed mutant strain. (Submitted to *American Zoologist*.)

Summerbell, D. (1974a). Interaction between the proximo-distal and antero-posterior co-ordinates of positional value during the specification of positional information in the early development of the chick limb-bud. *J. Embryol. exp. Morph.* **32**, 227–237.

Summerbell, D. (1974b). A quantitative analysis of the effect of excision of the AER from the chick limb bud. *J. Embryol. exp. Morph.* **32**, 651–660.

Summerbell, D. (1977). Regulation of deficiencies along the proximodistal axis of the chick wing bud: a quantitative analysis. *J. Embryol. exp. Morph.* **41**, 137–159.

Summerbell, D. (1979). The zone of polarizing activity: evidence for a role in normal chick limb morphogenesis. *J. Embrol. exp. Morph.* **50**, 217–233.

Summerbell, D., Lewis, J. H. and Wolpert, L. (1973). Positional information in chick limb morphogenesis. *Nature* **244**, 492–496.

Sussman, M. and Schindler, J. (1978). A possible mechanism for morphogenetic regulation in *Dictyostelium discoideum*. *Differentiation* **10**, 1–5.

Szabad, J., Simpson, P. and Nöthiger, R. (1979). Regeneration and compartments in *Drosophila*. *J. Embryol. exp. Morph.* **49**, 229–241.

Tank, P. W. and Holder, N. (1978). The effect of healing time on the proximodistal organization in the axolotl, *Ambystoma mexicanum*. *Dev. Biol.* **66**, 72–85.

Tardent, P. and Tardent, R. (1980). "Development and Cellular Biology of Coelenterates." Elsevier/North Holland, Amsterdam.

Tasaka, M. and Takeuchi, I. (1981). Role of cell sorting and pattern formation in *Dictiostelium discoideum*. *Differentiation* **18**, 191–196.

Tazima, Y. (1964). "The Genetics of the Silkworm". Logos, London.

Thoenen, H. and Barde, Y. A. (1980). Physiology of the Nerve Growth Factor. *Physiological Rev.* **60**, 1284–1334.

Tickle, C. (1981). The number of polarizing region cells required to specify additional digits in the developing chick wing. *Nature* **289**, 295–298.

Tickle, C., Summerbell, D. and Wolpert, L. (1975). Positional signalling and specification of digits in chick limb morphogenesis. *Nature* **254**, 199–202.

Town, C. D., Gross, J. D. and Kay, R. R. (1976). Cell differentiation without morphogenesis in *Dictyostelium discoideum*. *Nature* **262**, 717–719.

Trembley, A. (1744). Memoires pour servir a l'histoire d'un genre de polypes d'eau douce a bras en forme de cornes. Leyden.

Turing, A. (1952). The chemical basis of morphogenesis. *Phil. Trans. B.* **237**, 37–72.

van der Meer, J. (1978). Region specific cell differentiation during early insect development. Thesis, Nijmegen.

van der Meer, J. M., Ouveneel, W. J. (1974). Differentiation capacities of the dorsal metathoracic (haltere) disc of *Drosophila melanogaster*. II. Regeneration and duplication. *Wilhelm Roux Arch.* **174**, 361–373.

Velarde, M. G. and Normand, C. (1980). Convection. *Scientific American* **243**, No. 1 (July), 78–93.

Vogel, O. (1978). Pattern formation in the egg of the leafhopper *Euscelis plebejus* Fall. (Homoptera): Developmental capacities of fragment isolated from the polar egg region. *Dev. Biol.* **67**, 357.

Waddington, C. H., Needham, J. and Brachet, J. (1936). The activation of the evocator. *Proc. roy. Soc. B* **120**, 173–190.

Wardlaw, C. W. and Cutter, E. G. (1956). Experimental and analytical studies of pteridophytes. The effect of shallow incisions on organogenesis in *Dryopteris aristata* Druce. *Ann. Bot.* **20**, 39–57.

Webster, G. (1971). Morphogenesis and pattern formation in hydroids. *Biol. Rev.* **46**, 1–46.

Webster, G. and Wolpert, L. (1966). Studies on pattern regulation in hydra. *J. Embryol. exp. Morph.* **16**, 91–104.

Weiss, P. (1939). "The Principles of Development." Holt, New York.

Wieschaus, E. and Gehring, W. (1976). Clonal analysis of primordial disc cells in the early embryo of *Drosophila melanogaster*. *Dev. Biol.* **50**, 249–263.

Wigglesworth, V. B. (1940). Local and general factors in the development of "pattern" in *Rhodnius prolixus*. *J. exp. Biol.* **17**, 180–200.

Wigglesworth, V. B. (1954). Growth and regeneration in the tracheal system of an insect *Rhodnius prolixus* (Hemiptera). *Quart. J. micr. Sci.* **95**, 115–137.

Wigglesworth, V. B. (1959). The role of the epidermal cells in the "migration" of tracheoles in *Rhodnius prolixus* (hemiptera). *J. exp. Biol.* **36**, 632–640.

Wilby, O. K. and Webster, G. (1970a). Studies on the transmission of hypostome inhibition in hydra. *J. Embryol. exp. Morph.* **24**, 583–593.

Wilby, O. K. and Webster, G. (1970b). Experimental studies on axial polarity in hydra. *J. Embryol. exp. Morph.* **24**, 595–613.

Wilcox, M. and Smith, R. J. (1980). Compartments and distal outgrowth in the *Drosophila* imaginal wing disc. *Wilhelm Roux Arch.* **188**, 157–161.

Wilcox, M., Mitchison, G. J. and Smith, R. J. (1973). Pattern formation in the blue-green alga, *Anabaena*. I. Basic mechanisms. *J. Cell Sci.* **12**, 707–723.

Wolpert, L. (1969). Positional information and the spatial pattern of cellular differentiation. *J. theoret. Biol.* **25**, 1–47.

Wolpert, L. (1971). Positional information and pattern formation. *Curr. Top. Dev. Biol.* **6**, 183–224.

Wolpert, L. and Hornbruch, A. (1981). Positional signalling along the anteroposterior axis of the chick wing. The effect of multiple polarizing region grafts. *J. Embryol. exp. Morph.* **63**, 145–159.

Wolpert, L., Hicklin, J. and Hornbruch, A. (1971). Positional information and pattern regulation in regeneration of hydra. *Symp. Soc. exp. Biol.* **25**, 391–415.

Wolpert, L., Lewis, J. and Summerbell, D. (1975). Morphogenesis of the vertebrate limb. *In* "Cell Patterning". Ciba Foundation Symp. No. 29, pp. 95–118. Associated Scientific Publishers, Amsterdam.

Wolpert, L., Tickle, C. and Sampford, M. (1979). The effect of cell killing by X-irradiation on pattern formation in the chick limb. *J. Embryol. exp. Morph.* **50**, 175–198.

Wright, D. A. and Lawrence, P. A. (1981a). Regeneration of segment boundary in *Oncopeltus*. *Dev. Biol.* **85**, 317–327.

Wright, D. A. and Lawrence, P. A. (1981b). Regeneration of segment boundaries in *Oncopeltus*: cell lineage. *Dev. Biol.* **85**, 328–333.

Yajima, H. (1960). Studies on embryonic determination of the harlequin-fly, *Chironomous dorsalis*. *J. Embryol. exp. Morph.* **8**, 198–215.

Yajima, H. (1964). Studies on embryonic determination of the Harlequin-fly, *Chironomous dorsalis*. II. Effects of partial irradiation of the egg by ultraviolet light. *J. Embryol. exp. Morph.* **12**, 89–100.

Zwilling, E. (1961). Limb morphogenesis. *Adv. Morphogen.* **1**, 301–330.

Subject index

(Page number is given in *italic* if the subject is dealt with predominantly in a figure legend.)

A

Activation centre, in insects, 62, *63*
Activation of particular genes, 111–117
  of the Bithorax gene complex, 159–163
  computer programs and simulations, 200–205
Activator and inhibitor, 12
  strategies for isolation, 46, 47
Amphibians, *see also* vertebrate limbs,
  cilia pattern in the epidermis, *29*
  dorso-ventral organization, *150*
  organizing centres, 20
  somite determination, 166
Anastomosis, 25, 177
Apical ectodermal maintenance factor (AEMF), 118–123
Apical ectodermal ridge (AER), *98*, 103–105, 118–123
Autocatalysis,
  in gene activation, 111, *112*
  realization by an inhibition of an inhibition, 36, 110, 126
  requirement for pattern formation, 10–17
Auxin, 184

B

Bateson's rules, *86*
Bénard-instability, 129
Bicaudal mutation, *67*
Binary subdivisions, 76, *116*
Bithorax gene complex, 153, *154*, 159–163
  computer simulations, 200–205
Boundaries as organizing regions,
  in the eye primordia, 107
  in the imaginal disk, 82
  in the vertebrate limb, 95

Branching patterns, *see* net-like structures
Bristle-like pattern, *30, 31*
  computer simulations, 200
Brusselator, 34

C

Cyclic AMP, 42, 46
Cell aggregates,
  of hydra cells, 53
  of imaginal disc cells, 89
Chains of induction, 141
Chemotactic orientation,
  of *Dictyostelium*, 42, 45
  of nerve fibres, 46, 183
Chicken limbs, *see* vertebrate limbs
Chromatophores on squids, *31*
Cockroach legs,
  circumferential pattern, *144*
  proximo-distal pattern of segments, 90–93
  proximo-distal pattern within segments, *139*
Compartments, 78–81
  and the Bithorax gene complex, 153, *154*, 159–161
  and the circumferential leg pattern, 143–147
  formation during primary pattern interpretation, 156–159
  respecification, *85*, 131–135, 163
Compartment borders,
  as organizing regions, 78–89
  transgression of, 131–135
Competent zone, in amphibians, 95–97
Complete circle rule, 81, 83, 84, 105
Computer programs, 190–211

226

S

Sea urchin
  animal-vegetal axis, 59
  dorso-ventral organization, *21, 22,* 24
  polarity reversal in fragments, *22*
Secondary embryonic fields (subfields),
    see imaginal discs, insect append-
    ages, vertebrate limbs
Segmentation
  problem of, 168–170
  sequential addition of abdominal seg-
    ments, 163–165
Self-enhancement, *see* autocatalysis
Sequences of structures, 7, 56–57, 138–
    151
Shift of activator maxima,
  in elongation of net-like structures,
    173, 174
  in oscillating systems, 45
  in the depleted substrate model, 35
Size-regulation
  negative, in insects, 71, 72
  of an activator–inhibitor pattern, 39,
    40
  of morphogenetic gradients, 58–60
  of a mutual activation pattern, *126*
  of planarians, *136*
  of sequences of structures, 148
  of a slime mould slug, 41–44
Slime mould, *see Dictyostelium
    discoideum*
Smittia
  double abdomen formation, 66–69
  ligation experiments, *73*
Somites, formation of, 165–168
Source density, 37–39
  and head induction in hydra, *50–54*
  polarity determining influence, 37–39
Source–sink, gradient formation by, 58,
    59
Stomata, 29

Stolons of marine hydroids, 25–27
Stripes, formation of, 127–129
  computer simulations, 195–199
Supernumerary limbs, *93,* 100–102
Switches, biochemical, 108
Symmetrical pattern, *19,* 20
  in amphibians, 149, *150*
  in amphibian limbs, *96, 100, 101*
  in imaginal disc regeneration, *147*
  in insects, *67, 68, 130*
Symmetry breaking, 18, 39

T

Thalidomide, 122
Tracheae, *177*
Tumour angiogenesis factor, 185
Tumour growth and blood vessels, 185

U

Unspecific induction, 5, 23, 34, 134
  in insect embryos, 66–69

V

Venation, *see* net-like structures
Vertebrate limbs,
  antero-posterior organization, 95–97,
    103–105
  intercalation in the proximo-distal
    axis, 120–123
  progress-zone, 118, *119,* 165
  proximo-distal axis, 117–120
  regeneration of symmetrical limbs, 99,
    100
  supernumerary limbs, 100–102

Z

Zebra stripes, *129*
Zone of polarizing activity, 57, 103–105